AF780830

Advanced Antennas for Wireless Communication Systems

Advanced Antennas for Wireless Communication Systems

Editors

Trushit Upadhyaya
Hari Shankar Singh

MDPI

Basel • Beijing • Wuhan • Barcelona • Belgrade • Novi Sad • Cluj • Manchester

Editors

Trushit Upadhyaya
Electronics & Communication Engineering
Charotar University of Science & Technology (CHARUSAT)
Changa
India

Hari Shankar Singh
Electronics & Communication Engineering
Thapar Institute of Engineering & Technology
Patiala
India

Editorial Office
MDPI
St. Alban-Anlage 66
4052 Basel, Switzerland

This is a reprint of articles from the Special Issue published online in the open access journal *Micromachines* (ISSN 2072-666X) (available at: https://www.mdpi.com/journal/micromachines/special_issues/D2I49OR4RB).

For citation purposes, cite each article independently as indicated on the article page online and as indicated below:

Lastname, A.A.; Lastname, B.B. Article Title. *Journal Name* **Year**, *Volume Number*, Page Range.

ISBN 978-3-7258-0450-4 (Hbk)
ISBN 978-3-7258-0449-8 (PDF)
doi.org/10.3390/books978-3-7258-0449-8

© 2024 by the authors. Articles in this book are Open Access and distributed under the Creative Commons Attribution (CC BY) license. The book as a whole is distributed by MDPI under the terms and conditions of the Creative Commons Attribution-NonCommercial-NoDerivs (CC BY-NC-ND) license.

Contents

About the Editors

Trushit Upadhyaya

Trushit Upadhyaya (senior member, IEEE), received his B.E. degree from Gujarat University in 2004; his M.E. degree from the Institute of Telecommunication Research, University of South Australia, Adelaide, in 2007; and his Ph.D. degree in Satellite Antennas from the Charotar University of Science and Technology (CHARUSAT), Changa, in 2014. He was a Visiting Scientist with the Physical Research Laboratory (PRL), Ahmedabad. He is currently working as a Professor and the Head of the Department of Electronics and Communication Engineering, Faculty of Technology and Engineering, CHARUSAT. He has published several research articles in various peer reviewed international journals and conferences. His research interests include antenna system design and applied electromagnetics. He has carried out research and consultancy projects for Australian defense agencies, the Indian Space Research Organization (ISRO), the Gujarat Council on Science and Technology (GUJCOST), and private organizations. He has published and presented several research articles in his area of research in India and abroad, and he has delivered invited talks at various organizations across Gujarat.

Hari Shankar Singh

Hari Shankar Singh, holds a B.Tech. degree from the IEC College of Engineering and Technology, Greater Noida, Uttar Pradesh. He earned his bachelor's degree in 2011 and his Ph.D. from the Department of Electronics Engineering at the Indian Institute of Technology (Banaras Hindu University), Varanasi, India, in 2015. Dr. Singh embarked on his Ph.D. journey immediately after completing his bachelor's degree.

Currently, he serves as an Associate Professor in the Department of Electronics and Communication Engineering at Thapar Institute of Engineering and Technology, Patiala, Punjab, India. Before joining Thapar Institute, he worked as an Assistant Professor in the Department of Electronics and Communication Engineering at Jaypee University Anoopshar, Uttar Pradesh.

In recognition of his outstanding research, Dr. Singh was honored with the Best Ph.D. Thesis Award at the 3rd IEEE UPCON 2016 conference held in IIT (BHU), Varanasi, on December 9-11, 2016. His academic and professional journey has taken him to various countries, including Italy, Malaysia, Thailand, and Dubai, where he actively participated and presented his research papers. With a prolific research output, Dr. Singh has authored more than 55 research papers in reputable international journals and presented/published over 50 research papers at national and international conferences, symposiums, and workshops. He is also working on several projects that are granted by SERB; MeitY; DST; and Centre of Excellence, TIET Patiala. Dr. Singh is handling three research projects worth more than 1.5 Crores. Dr. Singh has also completed two projects worth 40 Lacs.

Preface

In recent years, wireless communication has undergone a transformative evolution, becoming a global network connecting billions of entities, from individuals to enterprises. This shift has enabled intelligent interconnections among various physical objects, including vehicles, smartphones, habitats, and more. At the core of this revolution is the Internet of Things (IoT), a crucial technology for the next generation of communication. The IoT seamlessly integrates cost-effective data collection and dissemination devices, such as sensors and RFID tags, enabling rapid communication across locations and times. The primary objective of this reprint is to present creative academic papers focusing on antennas for various communication system applications. Numerous studies have concentrated on improving microwave and radio frequency (RF) components, especially antenna systems, considered the foundation of wireless systems. Advanced communication antenna technology has significantly progressed in recent years, enhancing communication quality in home and business applications.

This reprint focuses on developing various antenna structures for wireless communication systems. A low-profile millimeter-wave broadband metasurface-based antenna is investigated. Characteristic mode analysis (CMA) guides the design and optimization, using a printed circuit board with differently sized coplanar patches to create broadside modes. Innovation is introduced by etching slots on the parasite patch, generating a new resonant mode to increase low-frequency bandwidth. The orthogonal modes of selected modes are shifted out of the operational region, reducing cross-polarization. A dual-band printable monopole antenna is presented, showing enhanced bandwidth and gain.

Moreover, the demand for fast communication has led to the construction of terahertz (THz) antennas with high data rates. The design of a THz Multiple-Input Multiple-Output (MIMO) antenna with a metamaterial is analyzed. The proposed two-port antenna design leverages a split-ring resonator patch that is complementary, achieving a 50 dB improvement in isolation and an ultra-broadband response.

In conclusion, a diverse range of antennas are showcased in this reprint, revealing their potential to revolutionize the next generation of wireless communication, as we navigate communication technologies' dynamic landscape, the integration of IoT, advancements in 5G, and the impending arrival of 6G open new frontiers for exploration. This compilation of academic contributions serves as a testament to ongoing efforts to push antenna design boundaries, ensuring optimal performance in the connected world. The journey from 5G to 6G is marked by innovation, shaping the future of wireless communication.

Trushit Upadhyaya and Hari Shankar Singh
Editors

micromachines

MDPI

Editorial

Editorial for Advanced Antennas for Wireless Communication Systems

Trushit Upadhyaya [1,*] and Hari Shankar Singh [2,3]

1 Department of Electronics and Communication Engineering, Chandubhai S. Patel Institute of Technology, Charotar University of Science and Technology (CHARUSAT), Changa 388421, India
2 Department of Electronics and Communication Engineering, Thapar Institute of Engineering and Technology, Patiala 147004, India; harishankar.singh@thapar.edu
3 Centre of Excellence in Emerging Materials (CEEMS), Thapar Institute of Engineering and Technology, Patiala 147004, India
* Correspondence: trushitupadhyaya.ec@charusat.ac.in

Citation: Upadhyaya, T.; Singh, H.S. Editorial for Advanced Antennas for Wireless Communication Systems. *Micromachines* **2024**, *15*, 115. https://doi.org/10.3390/mi15010115

Received: 25 December 2023
Accepted: 8 January 2024
Published: 10 January 2024

Copyright: © 2024 by the authors. Licensee MDPI, Basel, Switzerland. This article is an open access article distributed under the terms and conditions of the Creative Commons Attribution (CC BY) license (https://creativecommons.org/licenses/by/4.0/).

In recent years, there has been a significant expansion in wireless communication, evolving into a global network connecting billions of entities, including individuals and enterprises. This transformation has given rise to intelligent interconnections among various physical objects, such as vehicles, smartphones, habitats, and their occupants. A pivotal technology for the next generation of communication is the Internet of Things (IoT), incorporating cost-effective data collection and dissemination devices like sensors and RFID tags. These facilitate rapid communication between objects and individuals across any location and time, finding applications in remote healthcare, wearables, autonomous vehicles, wireless robots, and smart homes [1–3].

As the demand for application-oriented end-user apps continues to grow, the fifth generation of communication (5G) has replaced the fourth generation, requiring a thorough investigation of elements enhancing its applicability. This includes a discussion on antenna design and measures to improve antenna performance [4,5]. Notably, the implementation of 5G in the IoT and the emergence of the sixth generation (6G) have become areas of active research, with 6G expected to be implemented by 2030. The paper aims to cover the evolution from 5G to 6G, emphasizing the integration of IoT with both generations and addressing antenna design and performance enhancement for 5G [6–8].

The objective of this Special Issue is to present creative academic papers focusing on antennas for various communication system applications. As high-speed internet connections serve as vital conduits connecting multiple devices, user demands are escalating rapidly due to technological advancements in wireless communication systems. Numerous studies have concentrated on improving microwave and radio frequency (RF) components, especially antenna systems, considered the foundation of wireless systems. Advanced communication antenna technology has significantly progressed in recent years, enhancing communication quality in home and business applications.

Antenna design faces challenges in balancing cost-performance trade-offs and physical parameters while achieving ideal radiation performance. Research is exploring innovative technologies such as terahertz antennas, Massive-MIMO, UWB-MIMO, and MIMO to meet data rate requirements while keeping costs low. The Special Issue seeks designs with optimal performance.

Low-Profile Antenna Design (Contributions 1 and 5): In this work, a low-profile millimeter-wave broadband metasurface-based antenna is used. The characteristic mode analysis (CMA) is utilized in the design and optimization of the suggested antenna to direct the mode excitation. A small, printed circuit board with four sets of differently sized coplanar patches was used to create four neighboring broadside modes that are fed directly by a coaxial probe. Next, by etching slots on the parasite patch, a new resonant mode is produced to increase the low-frequency bandwidth. The radiation performance of the old modes remains mostly unchanged despite the addition of a new mode. Additionally, the orthogonal modes of the selected modes were moved out of the operational region by creating dual slots on the mid-patch fed by the coaxial probe. It lowers the degree of cross-polarization by shifting the selected modes' orthogonal modes outside of the working band. Further, in Contribution 5, a dual-band printable monopole antenna is created and displayed. With enhanced bandwidth and gain, the proposed antenna responds in a promising manner. With a strength of 3.7 dBi and 5.26 dBi, the antenna may radiate from 3.49 GHz to 3.82 GHz and from 4.83 GHz to 5.08 GHz, respectively, with a bandwidth of 9.09% and 5.06%. The newly created antenna is novel in that it has enough engineered resonating elements without requiring an additional reactive component. The substrate is a reasonably priced FR-4 laminate. For both resonances, this structure has an efficiency of more than 83%.

Moreover, Ultrabroadband Antenna Design (Contribution 2): The requirement for fast communication has made it possible to construct THz antennas with high data rates, speeds, and frequencies. This manuscript describes the design of a THz MIMO antenna with a metamaterial. The suggested two-port antenna design makes use of a split-ring resonator patch that is complementary. To demonstrate the improvement, the design outcomes are also contrasted with a straightforward patch antenna. A 50 dB improvement in isolation is shown in the design. It is possible to attain a broadband width of 8.3 THz using this complimentary split-ring resonator architecture. 90% of the bandwidth is utilized, indicating an ultrabroadband response.

In Contribution 10: For sub-6 GHz MIMO communication, a four-port dielectric resonator (DR)-based multiple-input multiple-output (MIMO) antenna is described. Dual-band resonance was achieved via an aperture feeding the dielectric resonator. The DRA operates at 3.3 GHz and 3.9 GHz in the modes $TE_{01\delta}$ and $TE_{10\delta}$, respectively. Without the need for an additional isolation mechanism, the built antenna provides port isolation of greater than 20 dB at the desired frequencies. The simulation calculation was performed using full-wave high-frequency simulation software. At 3.3 GHz and 3.9 GHz, respectively, the antenna's maximal gain and efficiency are 5.8 dBi and 6.2 dBi, and 88.6% and 90.2%, respectively. There are good MIMO diversity parameters for the suggested resonator.

In conclusion, a range of antennas has been developed within this Special Issue, showcasing their potential to become key candidates for revolutionizing the next generation of wireless communication.

Author Contributions: Writing—original draft preparation, H.S.S.; writing—review and editing, T.U. and H.S.S. All authors have read and agreed to the published version of the manuscript.

Funding: The authors are thankful to the Science and Engineering Research Board (SERB), Government of India, for providing financial support to carry out this work under a project (File No. CRG/2022/001256).

Conflicts of Interest: The author declares no conflicts of interest.

List of Contributions

1. Han, K.; Yan, Y.; Yan, Z.; Wang, C. Low-Profile Millimeter-Wave Metasurface-Based Antenna with Enhanced Bandwidth. *Micromachines* **2023**, *14*, 1403. https://doi.org/10.3390/mi14071403.
2. Armghan, A.; Aliqab, K.; Alsharari, M.; Alsalman, O.; Parmar, J.; Patel, S.K. Design and Development of Ultrabroadband, High-Gain, and High-Isolation THz MIMO Antenna with a Complementary Split-Ring Resonator Metamaterial. *Micromachines* **2023**, *14*, 1328. https://doi.org/10.3390/mi14071328.
3. An, W.; Tian, X.; Wang, J.; Wang, S. Low-Profile Dual-Polarized Double-Layer Microstrip Antenna for 5G and 5G Wi-Fi. *Micromachines* **2023**, *14*, 942. https://doi.org/10.3390/mi14050942.
4. Aliqab, K.; Lavadiya, S.; Alsharari, M.; Armghan, A.; Daher, M.G.; Patel, S.K. Design and Fabrication of a Low-Cost, Multiband and High Gain Square Tooth-Enabled

Metamaterial Superstrate Microstrip Patch Antenna. *Micromachines* **2023**, *14*, 163. https://doi.org/10.3390/mi14010163.
5. Xi, S.; Cai, J.; Shen, L.; Li, Q.; Liu, G. Dual-Band MIMO Antenna with Enhanced Isolation for 5G NR Application. *Micromachines* **2023**, *14*, 95. https://doi.org/10.3390/mi14010095.
6. Pandya, K.V. Low Profile Meandered Printed Monopole WiMAX/WLAN Antenna for Laptop Computer Applications. *Micromachines* **2022**, *13*, 2251. https://doi.org/10.3390/mi13122251.
7. Negri, E.; Fuscaldo, W.; Burghignoli, P.; Galli, A. A Leaky-Wave Analysis of Resonant Bessel-Beam Launchers: Design Criteria, Practical Examples, and Potential Applicationsat Microwave and Millimeter-Wave Frequencies. *Micromachines* **2022**, *13*, 2230. https://doi.org/10.3390/mi13122230.
8. Armghan, A.; Aliqab, K.; Sorathiya, V.; Alenezi, F.; Alsharari, M.; Ali, F. Design and Fabrication of the Split Ring Resonator Shaped Two-Element MIMO Antenna with Multiple-Band Operation for WiMAX/5G/Zigbee/Wi-Fi Applications. *Micromachines* **2022**, *13*, 2161. https://doi.org/10.3390/mi13122161.
9. Suresh, A.C.; Reddy, T.S.; Madhav, B.T.P.; Das, S.; Lavadiya, S.; Algarni, A.D.; El-Shafai, W. Investigations on Stub-Based UWB-MIMO Antennas to Enhance Isolation Using Characteristic Mode Analysis. *Micromachines* **2022**, *13*, 2088. https://doi.org/10.3390/mi13122088.
10. Patel, U.; Upadhyaya, T. Four-Port Dual-Band Multiple-Input Multiple-Output Dielectric Resonator Antenna for Sub-6 GHz 5G Communication Applications. *Micromachines* **2022**, *13*, 2022. https://doi.org/10.3390/mi13112022.

References

1. Zheng, J.; Ryl, D.S.; Bisdikian, C.; Mouftah, H.T. The internet of things. *IEEE Communications Magazine*, 10 November.
2. Andrews, J.G.; Buzzi, S.; Choi, W.; Hanly, S.V.; Lozano, A.; Soong, A.C.K.; Zhang, J.C. What will 5G be? *IEEE J. Sel. Areas Commun.* **2014**, *32*, 1065–1082. [CrossRef]
3. Agiwal, M.; Roy, A.; Saxena, N. Next generation 5G wireless networks: A comprehensive survey. *IEEE Commun. Surv.* **2016**, *18*, 1617–1655. [CrossRef]
4. Fuentes, M.; Carcel, J.L.; Dietrich, C.; Yu, L.; Garro, E.; Pauli, V.; Lazarakis, F.I.; Grondalen, O.; Bulakci, O.; Yu, J.; et al. 5G new radio evaluation against IMT—2020 key performance indicators. *IEEE Access* **2020**, *8*, 110880–110896. [CrossRef]
5. Kumar, S.; Dixit, A.S.; Malekar, R.R.; Raut, H.D.; Shevada, L.K. Fifth Generation Antennas: A Comprehensive Review of Design and Performance Enhancement Techniques. *IEEE Access* **2020**, *8*, 163568–163593. [CrossRef]
6. Rahman, E.A.; Aloi, D.N. Design of a 5G Sub-6 GHz Vehicular Cellular Antenna Element with Consistent Radiation Pattern Using Characteristic Mode Analysis. *Sensors* **2022**, *22*, 8862. [CrossRef] [PubMed]
7. Puri, V.; Singh, H.S. Design of an isolation improved MIMO antenna using metasurface based absorber for wireless applications. *Optik* **2022**, *259*, 168963. [CrossRef]
8. Di Paola, C.; Zhao, K.; Zhang, S.; Pedersen, G.F. SIW Multibeam Antenna Array at 30 GHz for 5G Mobile Devices. *IEEE Access* **2019**, *7*, 73157–73164. [CrossRef]

Disclaimer/Publisher's Note: The statements, opinions and data contained in all publications are solely those of the individual author(s) and contributor(s) and not of MDPI and/or the editor(s). MDPI and/or the editor(s) disclaim responsibility for any injury to people or property resulting from any ideas, methods, instructions or products referred to in the content.

micromachines

MDPI

Article

Low-Profile Millimeter-Wave Metasurface-Based Antenna with Enhanced Bandwidth

Ke Han [1,*], Yuchu Yan [1], Ze Yan [2] and Chongwei Wang [2]

[1] State Key Laboratory of Information Photonics and Optical Communications, Beijing University of Posts and Telecommunications, Haidian District, Beijing 100876, China; yanyuchu@bupt.edu.cn
[2] School of Electronic Engineering, Beijing University of Posts and Telecommunications, Haidian District, Beijing 100876, China; yanze2020@bupt.edu.cn (Z.Y.); chwzbj@bupt.edu.cn (C.W.)
* Correspondence: hanke@bupt.edu.cn

Abstract: A millimeter-wave broadband metasurface-based antenna with a low profile is proposed. In order to guide the mode excitation, the characteristic mode analysis (CMA) is used for the design and optimization of the proposed antenna. Four sets of coplanar patches with different dimensions on a thin printed circuit board are used to generate four adjacent broadside modes, which are directly fed by a coaxial probe. Then, to expand low-frequency bandwidth, a new resonant mode is introduced by etching slots on the parasite patch. Meanwhile, the extra mode introduced does not significantly change the radiation performance of the original modes. Moreover, dual slots are etched on the mid patch fed by the coaxial probe, which moves the orthogonal modes of the chosen modes out of the operating band to reduce cross-polarization levels. The proposed antenna realized 25.02 % (30–38.58 GHz) impedance bandwidth with dimensions of $1.423 \times 1.423 \times 0.029\lambda_0^3$ (λ_0 is the wavelength at 34 GHz in free space), and the realized gain in the band is 8.35–11.3 dB.

Keywords: broadband antenna; characteristic mode analysis (CMA); metasurface-based antenna; millimeter-wave antenna

Citation: Han, K.; Yan, Y.; Yan, Z.; Wang, C. Low-Profile Millimeter-Wave Metasurface-Based Antenna with Enhanced Bandwidth. *Micromachines* **2023**, *14*, 1403. https://doi.org/10.3390/mi14071403

Academic Editors: Trushit Upadhyaya and Hari Shankar Singh

Received: 14 June 2023
Revised: 6 July 2023
Accepted: 6 July 2023
Published: 10 July 2023

Copyright: © 2023 by the authors. Licensee MDPI, Basel, Switzerland. This article is an open access article distributed under the terms and conditions of the Creative Commons Attribution (CC BY) license (https://creativecommons.org/licenses/by/4.0/).

1. Introduction

Following the development of wireless communication technology, it has become deeply integrated into the daily life of humans. In recent years, the development of 5G has driven widespread demand for millimeter-wave technologies, which require greater bandwidth for high-speed data transmission. With the increasing use of millimeter-wave bands in wireless communications and the continuous increase in data transmission in communication scenarios, the ease of integration and the characteristics of the co-type make microstrip antennas promising for a wide range of applications. However, the disadvantage of the narrow operating band of microstrip antennas has become more and more prominent and has motivated the vigorous development of bandwidth expansion techniques for microstrip antennas.

A traditional microstrip antenna shows a narrow impedance bandwidth due to its single-resonance working mechanism. Etching slots [1] and loading short circuit pins or holes [2,3] are used to improve its performance. However, it shows low-level improvements in performance. Loading parasitic resonant patches [4–6] will increase cross-polarization levels and the analysis difficulty. Laminated patches [7,8] and air substrates [9] will introduce bulk to the antenna.

All the above techniques can broaden the bandwidth of the antenna well. However, these techniques make the structure of the antenna relatively complex, resulting in increasing analytical complexity. In order to implement broadband antennae with the advantages of low profile, low cost, and easy integration, metasurfaces have entered the research field. Early studies utilized composite periodic structures such as EBG (electromagnetic bandgap) and AMC (artificial magnetic conductor) to present an artificial impedance-controlled surface [10–12] for size reduction, wide bandwidth, and back radiation reduction. All these can be considered metasurface-based structures in a broad sense. On the one hand, metasurface is widely used in low reflection cross sections (RCS) [13–17] and filtering [18–21]. On the other hand, its multi-mode resonance characteristics show a prominent advantage in broadband [22–28], multi-band [29], and omnidirectional radiation [30–33] applications.

In addition, metasurface antennae not only retain the characteristics of low profile and easy integration, but also have the structure of multiple radiation patches, which can achieve broadside gain and broadband bandwidth that can exceed 50% [34–36], so it has attracted more and more attention. In our work, these methods are developed. The periodicity of the structure is broken to achieve a better bandwidth and frequency control capability. However, a traditional metasurface requires a multi-layer PCB feeding structure, which increases the antenna profile thickness. For millimeter-wave antennas, broadband antennas with more resonant modes, low profiles, and simple structures are needed to accommodate the increasing number of millimeter-wave mobile terminals.

It is a common design method for broadband metasurface antennas to excite multiple modes of metasurface for broadband operation [22,37]. With the help of CMA, the modal characteristics of the metasurface can be predicted to a certain extent, and the effects of structural changes can be revealed. In this way, it is possible to find a way to expand the bandwidth without significantly affecting the desired modes. In order to further increase the bandwidth without changing the overall size, targeted changes can be made to the characteristics mode of the metasurface by adjusting the patch size [29], patch segmentation [38,39], cutting angle [40], patch hollowing [34], and other methods. However, the coupling between the metasurface structures is very tight, and the desired modes are inevitably affected once the metasurface structure is drastically changed. For example, the sidelobe of Modal J_9 is suppressed by splitting the corner patches into four small patches, but the sidelobe level of the H-plane is increased [29]. Moreover, the segmented small-size patch has a nearly uniform effect on the desired radiation pattern, increasing the difficulty of adjusting a particular mode. In addition, complex feeding structures may also affect the characteristic mode of the metasurface, which increases the difficulty of designing the metasurface antenna. For example, dipole feeding [41] and slot coupling feeding [35,36] introduce new modes that can be used to expand the bandwidth of the antenna, but this puts forward higher requirements for the collaborative design of metasurface and feeding structure, and it will increase the profile.

In this paper, we propose a multi-mode resonant millimeter-wave metasurface-based antenna. Multiple patches are used to generate the characteristic broadside modes in adjacent frequency bands with coaxial probe excitation. Based on this, a new mode is introduced by etching slots on parasite patches, and an additional resonant point is added in the low-frequency region, which further enlarges the bandwidth of the antenna. Then, dual slots are etched on the mid patch fed by the coaxial probe, which moves the orthogonal modes of the chosen modes out of the operating band to reduce cross-polarization levels. Moreover, the etched slot does not significantly change the radiative properties of the original mode. Meanwhile, the metasurface-based structure consisting of multiple patches guarantees high antenna gain. Due to the simple feeding structure, the processing technique is straightforward. All this gives the antenna broadband, high gain, miniaturization, and a low profile.

2. Antenna Design

Figure 1 shows the configuration of the proposed antenna. The metasurface was printed on the top of a square single-layer PCB board with a thickness of 0.254 mm and a dimension of $W_g \times W_g$. The dielectric substrate was ROGERS RT/duroid 5880 ($\varepsilon_r = 2.2$). A total of 13 patches of 4 sizes were used, which were divided into C1, C2, C3, and C4 groups. Patch C1 is in the center, on which two slits with a width of 0.15 mm are etched. All patches except C1 are square. Two groups of 4 patches of different sizes are symmetrically arranged around C1. The rest are square patches with side length L_4. The spacing between C1 and

C2, C3, and C4 is S_{12}, S_{13}, and S_{14}, and the slots between C2 and C4, C3 and C4 is S_{24} and S_{34}, respectively. There is a metalized via under C1, which is used to feed, and its diameter is 0.3 mm. The dimensions are summarized in Table 1.

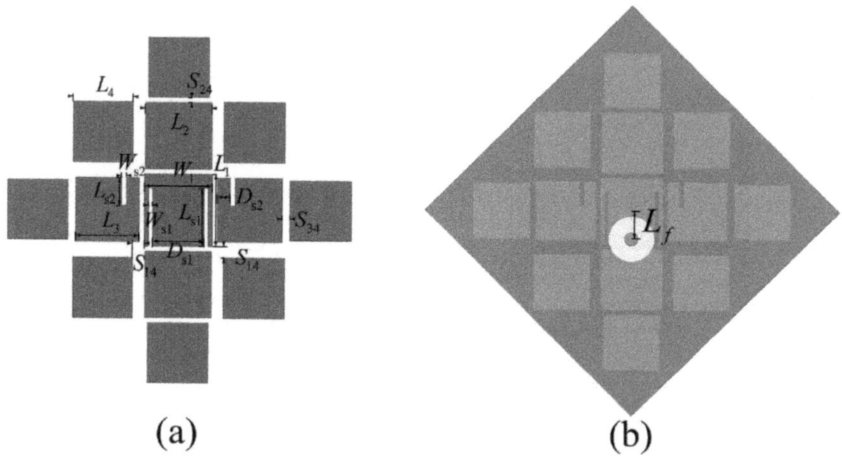

Figure 1. Configuration of the proposed antenna (**a**) top and (**b**) bottom.

Table 1. Summary of antenna geometry (unit: mm).

Par.	Val.	Par.	Val.	Par.	Val.
W_1	2.65	S_{13}	0.2	L_{s1}	2.3
L_1	2.94	S_{14}	0.45	L_{s2}	1.08
L_2	2.6	S_{24}	0.2	W_{s1}	0.15
L_3	2.53	S_{34}	0.3	W_{s2}	0.2
L_4	2.41	D_{s1}	2	L_f	1.2
S_{12}	0.2	D_{s2}	0.5		

The resonance behavior of source-free metasurface was characterized by CST. In this paper, the ground plane is infinite for CMA simulations, while in other simulations, the size of the ground plane is consistent with that of the dielectric substrate, which is $W_g \times W_g$. As shown in Figure 2, the lower surface of the dielectric substrate is set as the PEC boundary, and the remaining directions are open. When CST is used for characteristic mode analysis, the influence of feeding structure is not considered, which means CMA is carried out for metasurface without feeding structure. Five adjacent broadside characteristic modes are selected. The modal significance is shown in Figure 3, and the resonant frequencies are 28.864, 32.524, 35.656, 37.264, and 39.512 GHz, respectively. The corresponding modal current and radiation patterns at the resonant frequency are shown in Figures 4 and 5. The modal current of Mode J_1 is mainly concentrated around the slots of C3 patches, which is called slot mode. The mode currents of the other four modes are concentrated in the patch sets of C1, C2, C3, and C4, respectively, which are dominated by the corresponding patch sets.

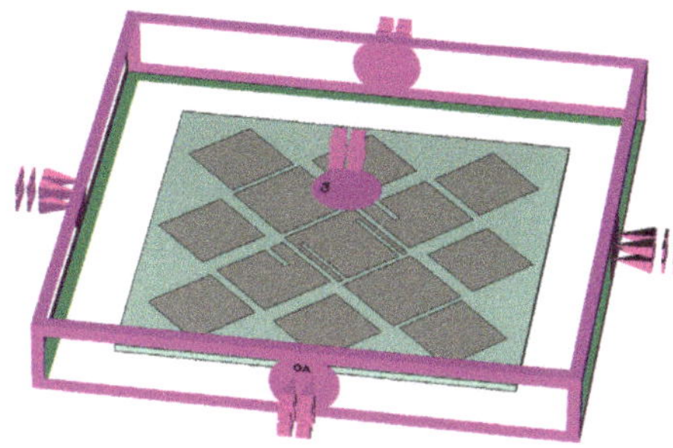

Figure 2. Geometrical modeling and boundary setup.

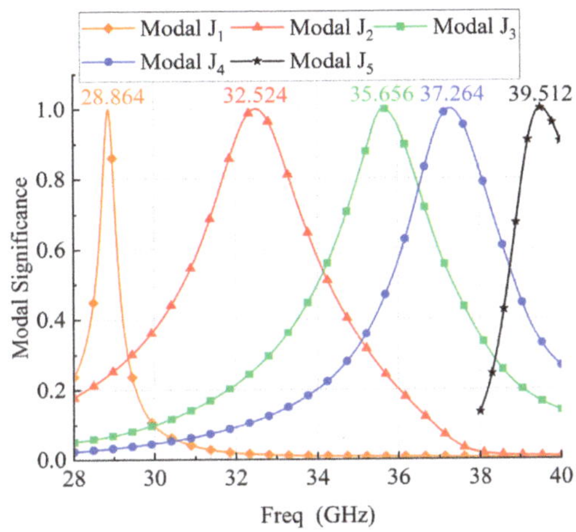

Figure 3. Modal significances of the proposed metasurface-based antenna.

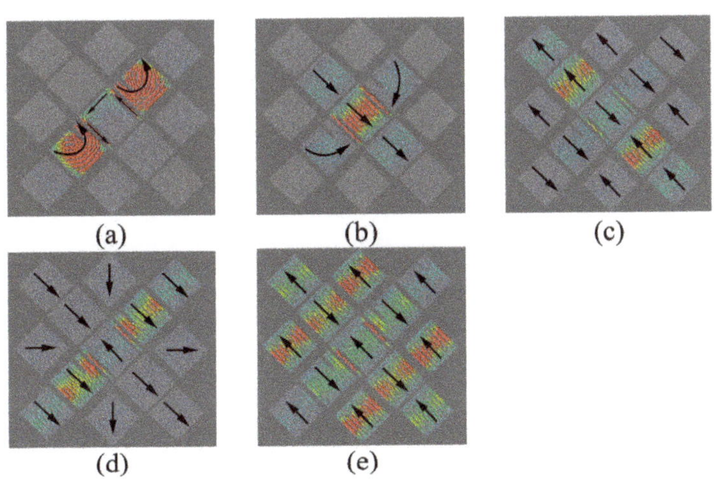

Figure 4. Modal currents at resonant frequency for (**a**) Modal J_1; (**b**) Modal J_2; (**c**) Modal J_3; (**d**) Modal J_4.

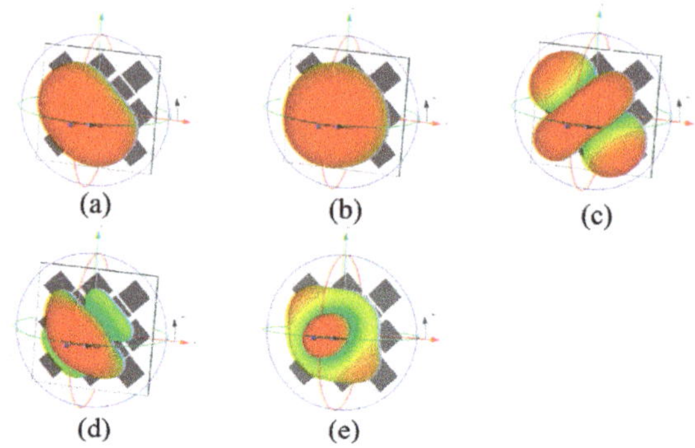

Figure 5. Modal radiation pattern at resonant frequency for (**a**) Modal J_1; (**b**) Modal J_2; (**c**) Modal J_3; (**d**) Modal J_4; and (**e**) Modal J_5.

The mainly modal currents are polarized along -45° and can be excited simultaneously by a coaxial probe. The modal significances of Modal J_2 gradually decreases after 32.5 GHz, and the contribution of the radiation mode provided by Modal J_2 to the total radiation mode of the antenna also gradually decreases. After Modal J_2 degenerates to a non-significant mode (MS $\leq$ 0.707), there is no other mode that can provide the required radiation mode. Therefore, the antenna gain in this frequency range will gradually decrease. Until the contribution of Modal J_3 and Modal J_4 to the total radiation pattern of the antenna cancels out the decreasing trend due to the attenuation of the radiation pattern provided by Modal J_2. In other words, the half-power bandwidth of Modal J_2 and Modal J_3 does not completely cover the frequency interval between the resonance points of the two modes, where the antenna gain deteriorates. Mode J_3 and Mode J_4 were both significant modes (MS $\geq$ 0.707) between their resonant frequency points, which makes these two modes present as mixed modes.

Modal J_5, which is dominated by the C4 patch group and has a resonant frequency of 39.5 GHz, is not in the operating frequency band. Modal J_5 forms a resonance point with the Modal J_4 that degenerates into an insignificant pattern. As frequencies close to 39.5 GHz, the Modal J_5 excitation level gradually increases and contributes to the dominant radiative features, which makes the antenna gain increase. Thus, the antenna gain still shows an increasing trend beyond the half-power bandwidth of the Modal J_4.

Figure 6 shows the characteristic mode analysis results of the grooveless metasurface-based antenna, which is called Antenna 2 (Figure 7b). Compared to Figure 3, the resonant frequencies of Modal J_2 and Modal J_4 change slightly after etched slots, while the radiative properties of the other modes do not change significantly. The slots etched on C3 introduce a new resonant mode, Modal J_1, in the low-frequency region, which further broadens the frequency band of the metasurface-based antenna. The etched slots on C1 cause the orthorhombic modes of the selected mode to move out of the band, optimizing the cross-polarization level. Modal currents of Modal J_1 are mainly distributed on both sides of the slots. The modal current flows around the slots on the C3 patch, consisting of two polarization currents polarized along and perpendicular to the slots, and the polarization purity of the modal current is not very high. When a coaxial probe is used for feeding, the electric field emanating from the feeding point cannot excite it to the maximum extent, and the antenna gain will be low when this mode dominates. The resonant frequency is affected by the dimensions of the slots, which can be adjusted to shift the mode into the band.

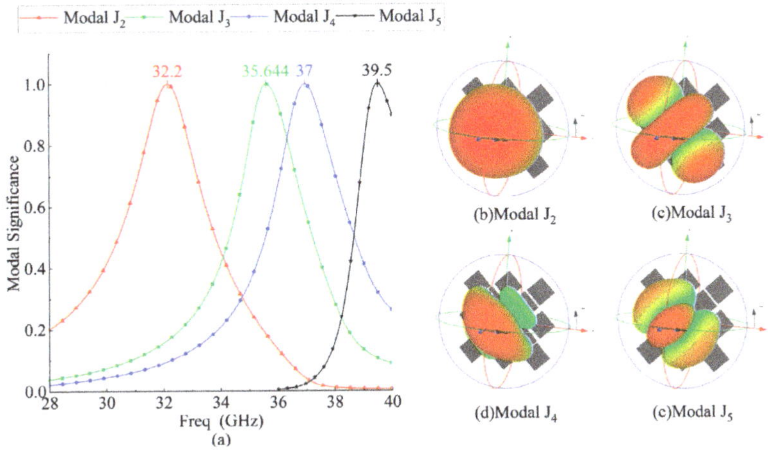

Figure 6. (**a**) Modal significances of the metasurface without slots; modal radiation pattern of the metasurface without slots for (**b**) Modal J_2; (**c**) Modal J_3; (**d**) Modal J_4; and (**e**) Modal J_5.

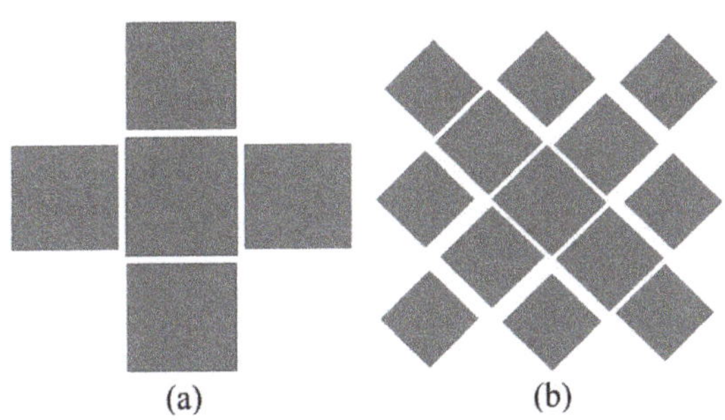

Figure 7. Configuration of the reference antenna for (**a**) antenna 1 and (**b**) antenna 2.

To evaluate the effectiveness of this strategy, the performance of this antenna is compared with that of two reference antennas (Figure 7). Simulated $|S_{11}|$ and broadside gain is shown in Figure 8. The slots etched on the C3 patch dominates the Modal J_1, introduces a new resonant point in the low-frequency region, and widens the impedance bandwidth of the metasurface-based antenna by 0.56 GHz. Since the polarization purity of the Modal J_1 is not high, the gain of the antenna is low with respect to the other frequency points when this mode dominates. The Modal J_2 is dominated by the C1 patch, and the etching slots essentially do not change its resonance frequency, only the impedance matching. Modal J_5, which is dominated by the C4 patches, introduces a new resonance point in the high-frequency region, further expanding the bandwidth. Since the modal currents on C4 patches are in phase with the primary polarization currents, the gain of the antenna is enhanced. When Modal J_3 and Modal J_4 dominate, the modal current on C4 patches is in phase with the main polarization current, which improves the gain deterioration in Antenna 1. However, the currents on the C1 patch are out of phase with the main polarization current, and the etched slots enhance the out-of-phase currents on the C1 patch and weaken the gain when Modal J_2 and Modal J_3 dominate. Hence the gain of Pro.Ant deteriorates again. This gain fluctuation is acceptable, considering the bandwidth gain from the etching slots.

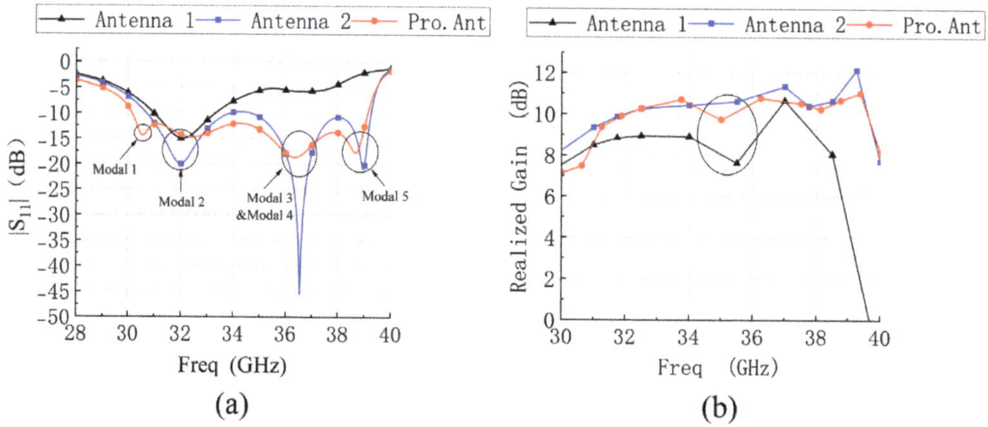

Figure 8. (**a**) Simulated $|S_{11}|$ and (**b**) broadside realized gains of reference antenna 1, antenna 2, and proposed antenna.

After the introduction of the new resonance point, there is a peak impedance in the low-frequency region, as shown in Figure 9. The real part of the peak impedance increases by 10 Ω, and the reactance increases from −9.7 to −5.6. The impedance matching with the 50 Ω port is significantly improved, while the impedance of other frequencies also approaches the impedance of 50 + j0 Ω.

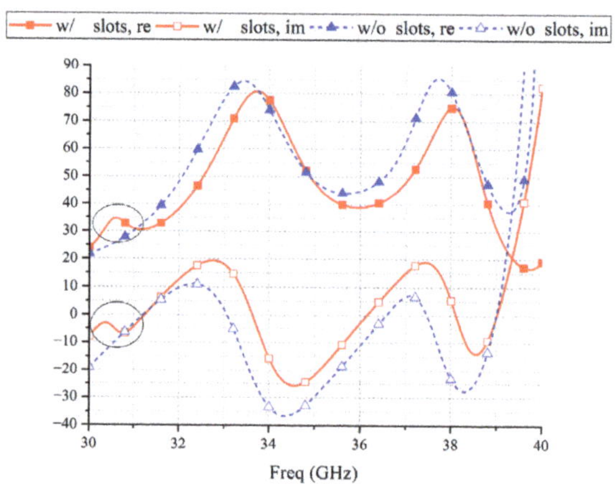

Figure 9. The influences of slots on impedance.

3. Parametric Study

The influences of L_2, L_3, L_4, and L_{s2} on $|S_{11}|$ are shown in Figure 10. When Modal J_2, Modal J_3, and Modal J_5 are dominant, there is a strong polarization current across the C2 patches, so these modes are affected to varying degrees when L_2 is varied. The polarization current of Modal J_4 is mainly concentrated on the C3 patches. Since Modal J_4 is mixed to varying degrees with Modal J_3 and Modal J_5, the impedance matching of the two resonant points in the high-frequency region can be tuned by adjusting L_3, and the resonant frequency does not change significantly.

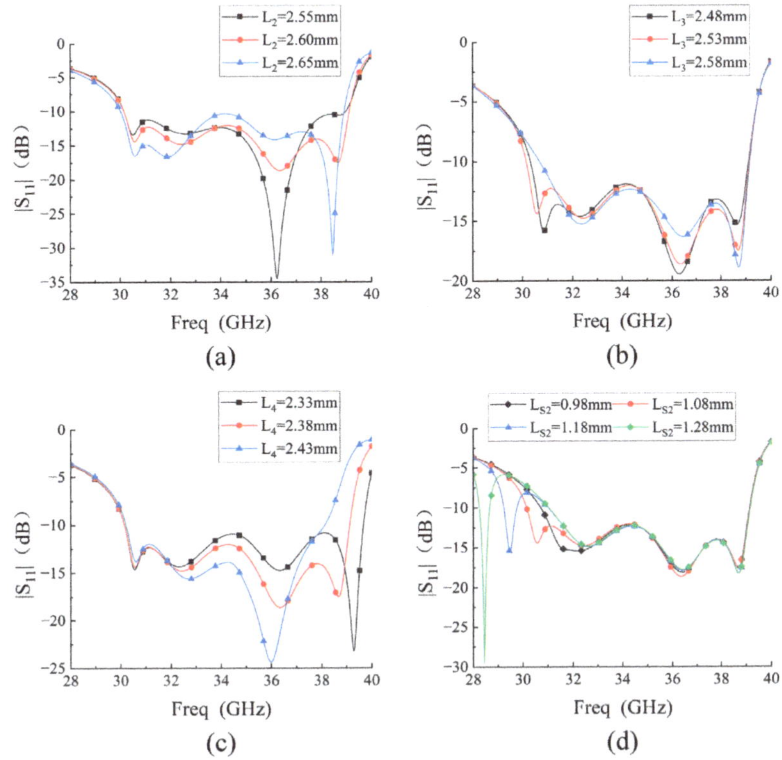

Figure 10. Influences of (**a**) L_2, (**b**) L_3, (**c**) L_4, and (**d**) L_{s2} on $|S_{11}|$.

As the length of L_4 decreases, the resonant frequency of the fourth resonant point gradually shifts to a higher frequency, while the frequencies of other resonant points hardly change. As the length of L_{s2} increases, so does the length of the current path flowing around the slots, and thus the resonant frequency shifts to lower frequencies. When the slot length L_{s2} is shortened to 0.98 mm, the resonant frequency of the slot mode is very close to that of mode 1. If the length of L_{s2} is further shortened, the resonant point generated by the slot mode can no longer be observed independently. The variation of the slot length does not affect the characteristics of the other modes.

Considering bandwidth and impedance matching, the four parameters are set as $L_2 = 2.60$ mm, $L_3 = 2.53$ mm, $L_4 = 2.38$ mm, and $L_{s2} = 1.08$ mm.

4. Measured Results

A prototype sample of the proposed metasurface-based antenna is shown in Figure 11. The metasurface-based antenna is printed by a single-layer PCB with a copper foil thickness of 0.018 mm. It is fed by a 3 × 3 × 3 mm^3 RF coaxial connector with a characteristic impedance of 50 Ω (the coaxial probe is not inserted into the dielectric substrate, only welded to the bottom pad). The antenna is measured by an Agilent 5230Avector network analyzer and the standard anechoic chamber, as Figure 12 shows.

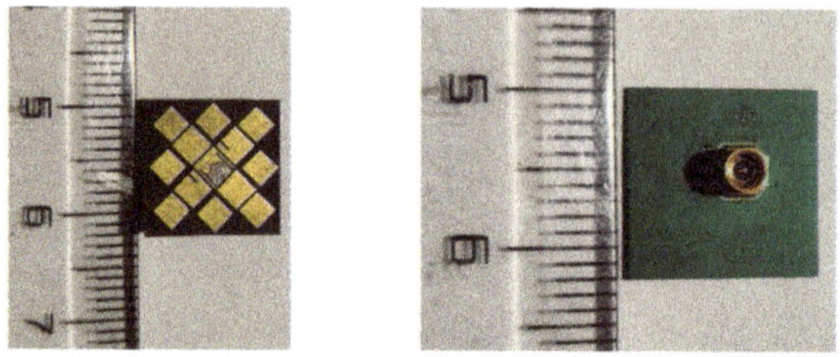

(a) (b)

Figure 11. Prototype of the proposed metasurface-based antenna's (**a**) top view and (**b**) bottom view.

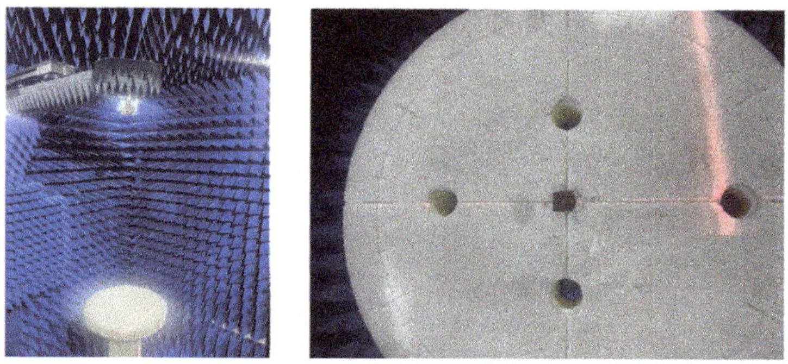

Figure 12. Far-field measurement environment.

Figures 13 and 14 show the measured results of the antenna impedance matching, peak gain, and radiation pattern. Due to welding, thick, irregularly shaped solder is deposited at the feeding point and edges of the C1 patch at the top of the antenna, which causes a slight change in the size of the C1 patch, resulting in a difference in impedance matching and resonant frequency. The measured impedance bandwidth of the metasurface-based antenna is 25.02% (30–38.58 GHz), and the realized gain in the band is 8.35–11.3 dB. The performance of the metasurface-based antenna is in line with expectations.

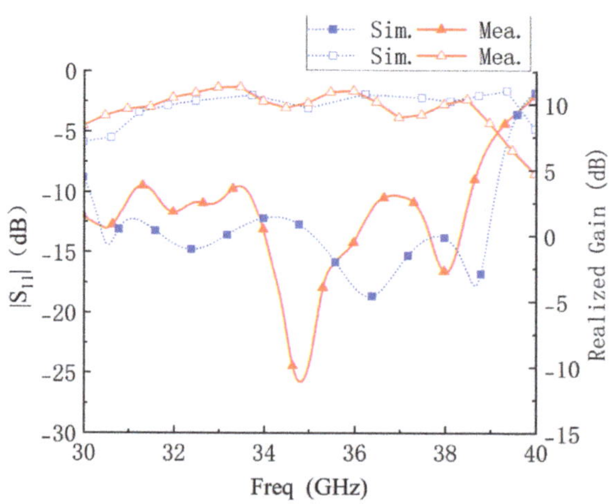

Figure 13. Simulated and measured $|S_{11}|$ and broadside gain.

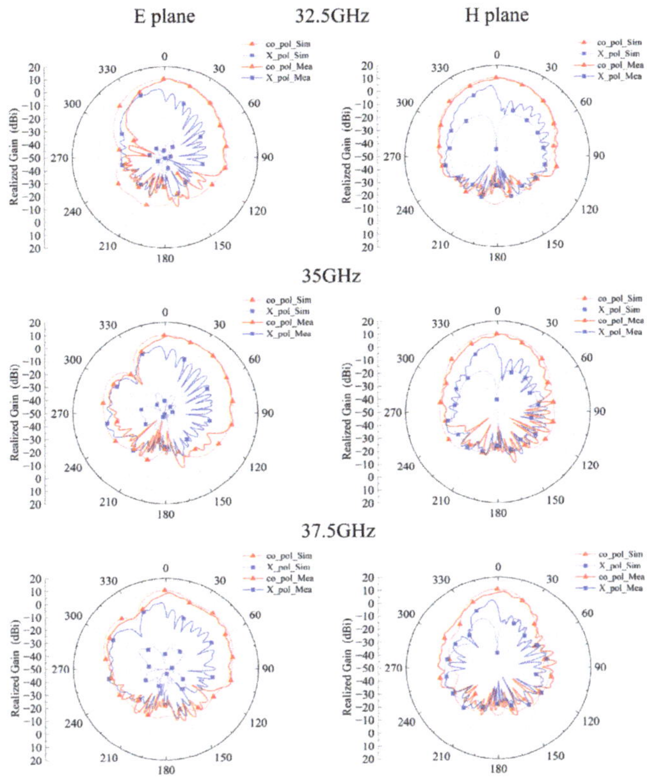

Figure 14. Simulated and measured radiation patterns at 32.5, 35, and 37.5 GHz.

As shown in Figure 15, the radiation efficiency of the antenna in the high-frequency region decreases gradually, and the in-band efficiency stays above 75%. The rapid decrease in radiation efficiency of the proposed antenna is due to the effects of the SSMP connector. Also, the effects of the measured equipment could not be completely ruled out as a possible cause.

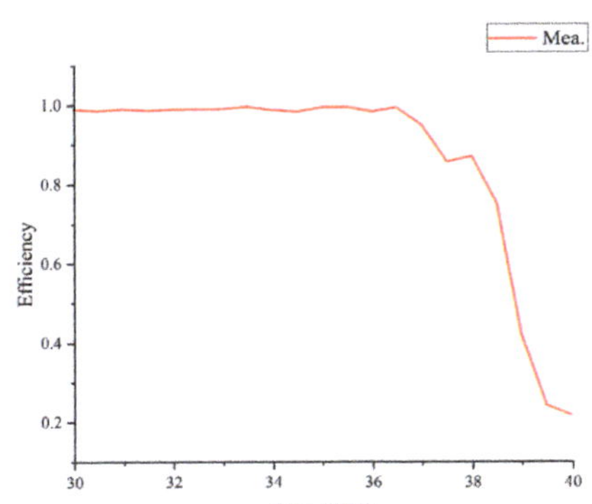

Figure 15. Measured radiation efficiency of the proposed antenna.

5. Conclusions

In this paper, we propose a low-profile broadband metasurface-based antenna with five resonant modes. CMA is used to guide the design of the proposed antenna. A new resonant mode is introduced by etching slots, and a coaxial probe is used to excite all modes, resulting in a wide bandwidth. The etched slots expand the bandwidth without significantly affecting the original mode while reducing the cross-polarization level. The proposed antenna realized 25.02% (30–38.58 GHz) impedance bandwidth, and the realized gain in the band is 8.35–11.3 dB. The antenna possesses the advantages of having a small size, large bandwidth, and practical value.

Author Contributions: Validation, Y.Y. and C.W.; Investigation, Z.Y.; Writing—original draft, Y.Y.; Writing—review & editing, K.H.; Project administration, K.H. All authors have read and agreed to the published version of the manuscript.

Funding: This research was funded by the Beijing Municipal Natural Science Foundation (Grant No. 4222072) and the Fund of State Key Laboratory of Information Photonics and Optical Communications (Beijing University of Posts and Telecommunications) (No. IPOC2021ZT20), P. R. China.

Data Availability Statement: Data sharing not applicable.

Conflicts of Interest: The authors declare no conflict of interest.

References

1. Radavaram, S.; Pour, M. Wideband Radiation Reconfigurable Microstrip Patch Antenna Loaded with Two Inverted U-Slots. *IEEE Trans. Antennas Propag.* **2018**, *67*, 1501–1508. [CrossRef]
2. Zaker, R.; Kheirdoost, A. Bandwidth and Isolation Improvement of Highly Coupled Printed Array Antenna Using Multiple Shorting Posts. *IEEE Trans. Antennas Propag.* **2021**, *69*, 7987–7992. [CrossRef]
3. Liu, J.; Zheng, S.; Li, Y.; Long, Y. Broadband Monopolar Microstrip Patch Antenna with Shorting Vias and Coupled Ring. *IEEE Antennas Wirel. Propag. Lett.* **2013**, *13*, 39–42. [CrossRef]
4. Yang, D.; Zhai, H.; Guo, C.; Li, H. A Compact Single-Layer Wideband Microstrip Antenna with Filtering Performance. *IEEE Antennas Wirel. Propag. Lett.* **2020**, *19*, 801–805. [CrossRef]
5. Midya, M.; Bhattacharjee, S.; Mitra, M. Broadband Circularly Polarized Planar Monopole Antenna With G-Shaped Parasitic Strip. *IEEE Antennas Wirel. Propag. Lett.* **2019**, *18*, 581–585. [CrossRef]
6. Nishiyama, E.; Aikawa, M. Wide-band and high-gain microstrip antenna with thick parasitic patch substrate. *Proc. IEEE Antennas Propag. Soc. Int. Symp.* **2004**, *1*, 273–276. [CrossRef]
7. Ding, C.; Liu, L.; Luk, K.-M. An Optically Transparent Dual-Polarized Stacked Patch Antenna with Metal-Mesh Films. *IEEE Antennas Wirel. Propag. Lett.* **2019**, *18*, 1981–1985.
8. Yang, W.; Zhou, J.; Yu, Z.; Li, L. Single-Fed Low Profile Broadband Circularly Polarized Stacked Patch Antenna. *IEEE Trans. Antennas Propag.* **2014**, *62*, 5406–5410. [CrossRef]
9. Sarkar, T.; Ghosh, A.; Singh, L.K.; Chattopadhyay, S.; Sim, C.-Y. DGS-Integrated Air-Loaded Wideband Microstrip Antenna for *X*- and *Ku*-Band. *IEEE Antennas Wirel. Propag. Lett.* **2019**, *19*, 114–118. [CrossRef]
10. Mosallaei, H.; Sarabandi, K. Novel artificial reactive impedance surface for miniaturized wideband planar antenna design: Concept and characterization. In Proceedings of the IEEE Antennas and Propagation Society International Symposium. Digest. Held in conjunc-tion with: USNC/CNC/URSI North American Radio Sci. Meeting (Cat. No. 03CH37450), Columbus, OH, USA, 22–27 June 2003; IEEE: Piscataway, NJ, USA, 2003; Volume 2, pp. 403–406.
11. De Cos, M.E.; Álvarez, Y.; Las-Heras, F. Enhancing patch antenna bandwidth by means of uniplanar EBG-AMC. *Microw. Opt. Technol. Lett.* **2011**, *53*, 1372–1377. [CrossRef]
12. Hadarig, R.C.; De Cos, M.E.; Las-Heras, F. Microstrip patch antenna bandwidth enhancement using AMC/EBG structures. *Int. J. Antennas Propag.* **2012**, *2012*, 843754. [CrossRef]
13. De Cos, M.E.; Alvarez-Lopez, Y.; Andres, F.L.H. On the influence of coupling AMC resonances for RCS reduction in the SHF band. *Prog. Electromagn. Res.* **2011**, *117*, 103–119. [CrossRef]
14. Li, T.; Yang, H.; Li, Q.; Jidi, L.; Cao, X.; Gao, J. Broadband Low-RCS and High-Gain Microstrip Antenna Based on Concentric Ring-Type Metasurface. *IEEE Trans. Antennas Propag.* **2021**, *69*, 5325–5334. [CrossRef]
15. Zheng, Q.; Guo, C.; Ding, J.; Vandenbosch, G.A.E. A Broadband Low-RCS Metasurface for CP Patch Antennas. *IEEE Trans. Antennas Propag.* **2020**, *69*, 3529–3534. [CrossRef]
16. Zhao, Y.; Cao, X.; Gao, J.; Yao, X.; Liu, X. A Low-RCS and High-Gain Slot Antenna Using Broadband Metasurface. *IEEE Antennas Wirel. Propag. Lett.* **2015**, *15*, 290–293. [CrossRef]
17. Zhao, Y.; Cao, X.; Gao, J.; Yao, X.; Liu, T.; Li, W.; Li, S. Broadband Low-RCS Metasurface and Its Application on Antenna. *IEEE Trans. Antennas Propag.* **2016**, *64*, 2954–2962. [CrossRef]

18. Pan, Y.M.; Hu, P.F.; Zhang, X.Y.; Zheng, S.Y. A Low-Profile High-Gain and Wideband Filtering Antenna with Metasur-face. *IEEE Trans. Antennas Propag.* **2016**, *64*, 2010–2016. [CrossRef]
19. Yang, W.; Chen, S.; Xue, Q.; Che, W.; Shen, G.; Feng, W. Novel Filtering Method Based on Metasurface Antenna and Its Application for Wideband High-Gain Filtering Antenna with Low Profile. *IEEE Trans. Antennas Propag.* **2018**, *67*, 1535–1544. [CrossRef]
20. Chen, C.; Chen, J.; Zhou, J.; Wen, L.; Hong, W. Millimeter-Wave Filtering Metasurface Antenna Array with Printed RGW Technology. *IEEE Antennas Wirel. Propag. Lett.* **2023**, *22*, 1622–1626. [CrossRef]
21. Guo, J.; Chen, Y.; Yang, D.; Ma, B.; Liu, S.; Pan, J. Design of a Circuit-Free Filtering Metasurface Antenna Using Characteristic Mode Analysis. *IEEE Trans. Antennas Propag.* **2022**, *70*, 12322–12327. [CrossRef]
22. Xue, M.; Wan, W.; Wang, Q.; Cao, L. Low-Profile Millimeter-Wave Broadband Metasurface Antenna with Four Resonances. *IEEE Antennas Wirel. Propag. Lett.* **2021**, *20*, 463–467. [CrossRef]
23. Nasser, S.S.S.; Liu, W.; Chen, Z.N. Wide Bandwidth and Enhanced Gain of a Low-Profile Dipole Antenna Achieved by Integrated Suspended Metasurface. *IEEE Trans. Antennas Propag.* **2018**, *66*, 1540–1544. [CrossRef]
24. Wang, J.; Wang, W.; Liu, A.; Guo, M.; Wei, Z. Broadband Metamaterial-Based Dual-Polarized Patch Antenna with High Isolation and Low Cross Polarization. *IEEE Trans. Antennas Propag.* **2021**, *69*, 7941–7946. [CrossRef]
25. Liu, W.E.I.; Chen, Z.N.; Qing, X.; Shi, J.; Lin, F.H. Miniaturized Wideband Metasurface Antennas. *IEEE Trans. Antennas Propag.* **2017**, *65*, 7345–7349. [CrossRef]
26. Feng, G.; Chen, L.; Xue, X.; Shi, X. Broadband Surface-Wave Antenna with a Novel Nonuniform Tapered Metasurface. *IEEE Antennas Wirel. Propag. Lett.* **2017**, *16*, 2902–2905. [CrossRef]
27. Liu, P.; Jiang, W.; Hu, W.; Sun, S.-Y.; Gong, S.-X. Wideband Multimode Filtering Circular Patch Antenna. *IEEE Trans. Actions Antennas Propag.* **2021**, *69*, 7249–7259. [CrossRef]
28. Cao, T.N.; Nguyen, M.T.; Phan, H.L.; Nguyen, D.D.; Vu, D.L.; Nguyen, T.Q.H.; Kim, J.-M. Millimeter-Wave Broadband MIMO Antenna Using Metasurfaces for 5G Cellular Networks. *Int. J. RF Microw. Comput. Aided Eng.* **2023**, *2023*, 9938824. [CrossRef]
29. Li, T.; Chen, Z.N. A Dual-Band Metasurface Antenna Using Characteristic Mode Analysis. *IEEE Trans. Antennas Propag.* **2018**, *66*, 5620–5624. [CrossRef]
30. Liu, S.; Yang, D.; Chen, Y.; Sun, K.; Zhang, X.; Xiang, Y. Design of Single-Layer Broadband Omnidirectional Metasurface Antenna Under Single Mode Resonance. *IEEE Trans. Antennas Propag.* **2021**, *69*, 6947–6952. [CrossRef]
31. Liu, S.; Yang, D.; Chen, Y.; Zhang, X.; Xiang, Y. Compatible Integration of Circularly Polarized Omnidirectional Metasurface Antenna with Solar Cells. *IEEE Trans. Antennas Propag.* **2019**, *68*, 4155–4160. [CrossRef]
32. Liu, S.; Yang, D.; Wu, L.; Sun, K.; Hu, J.; Chen, Y. A Dual-Polarized Omnidirectional Metasurface Antenna Designed via Characteristic Mode Analysis. *IEEE Antennas Wirel. Propag. Lett.* **2022**, *22*, 1010–1014. [CrossRef]
33. Xu, Y.; Wen, S.; Dong, Y. Vertically Polarized Loop-Fed Slot Antenna with Top-Loading Metasurface for Omnidirectional LTE Base Station Application. *IEEE Antennas Wirel. Propag. Lett.* **2021**, *20*, 2397–2401. [CrossRef]
34. Chen, D.; Yang, W.; Che, W.; Xue, Q. Broadband Stable-Gain Multiresonance Antenna Using Nonperiodic Square-Ring Metasurface. *IEEE Antennas Wirel. Propag. Lett.* **2019**, *18*, 1537–1541. [CrossRef]
35. Liu, S.; Yang, D.; Pan, J. A Low-Profile Broadband Dual-Circularly-Polarized Metasurface Antenna. *IEEE Antennas Wirel. Propag. Lett.* **2019**, *18*, 1395–1399. [CrossRef]
36. Liu, S.; Yang, D.; Chen, Y.; Sun, K.; Zhang, X.; Xiang, Y. Low-Profile Broadband Metasurface Antenna Under Multimode Resonance. *IEEE Antennas Wirel. Propag. Lett.* **2021**, *20*, 1696–1700. [CrossRef]
37. Lin, F.H.; Chen, Z.N. Probe-fed broadband low-profile metasurface antennas using characteristic mode analysis. In Proceedings of the 2017 Sixth Asia-Pacific Conference on Antennas and Propagation (APCAP), Xi'an, China, 16–19 October 2017.
38. Li, T.; Chen, Z.N. Design of dual-band metasurface antenna. In Proceedings of the 2018 International Workshop on Antenna Technology (iWAT), Nanjing, China, 5–7 March 2018.
39. Lin, F.H.; Chen, Z.N. A Method of Suppressing Higher Order Modes for Improving Radiation Performance of Metasurface Multiport Antennas Using Characteristic Mode Analysis. *IEEE Trans. Antennas Propag.* **2018**, *66*, 1894–1902. [CrossRef]
40. Liu, J.; Weng, Z.; Zhang, Z.-Q.; Qiu, Y.; Zhang, Y.-X.; Jiao, Y.-C. A Wideband Pattern Diversity Antenna with a Low Profile Based on Metasurface. *IEEE Antennas Wirel. Propag. Lett.* **2021**, *20*, 303–307. [CrossRef]
41. Lin, F.H.; Chen, Z.N. Truncated Impedance Sheet Model for Low-Profile Broadband Nonresonant-Cell Metasurface Antennas Using Characteristic Mode Analysis. *IEEE Trans. Antennas Propag.* **2018**, *66*, 5043–5051. [CrossRef]

Disclaimer/Publisher's Note: The statements, opinions and data contained in all publications are solely those of the individual author(s) and contributor(s) and not of MDPI and/or the editor(s). MDPI and/or the editor(s) disclaim responsibility for any injury to people or property resulting from any ideas, methods, instructions or products referred to in the content.

micromachines

MDPI

Article

Design and Development of Ultrabroadband, High-Gain, and High-Isolation THz MIMO Antenna with a Complementary Split-Ring Resonator Metamaterial

Ammar Armghan [1,*], Khaled Aliqab [1,*], Meshari Alsharari [1], Osamah Alsalman [2], Juveriya Parmar [3,4] and Shobhit K. Patel [4]

1 Department of Electrical Engineering, College of Engineering, Jouf University, Sakaka 72388, Saudi Arabia
2 Department of Electrical Engineering, College of Engineering, King Saud University, P.O. Box 800, Riyadh 11421, Saudi Arabia
3 Department of Mechanical and Materials Engineering, University of Nebraska-Lincoln, 1400 R St. Nebraska, Lincoln, NE 68588, USA
4 Department of Computer Engineering, Marwadi University, Rajkot 360003, India
* Correspondence: aarmghan@ju.edu.sa (A.A.); kmaliqab@ju.edu.sa (K.A.)

Abstract: The need for high-speed communication has created a way to design THz antennas that operate at high frequencies, speeds, and data rates. In this manuscript, a THz MIMO antenna is designed using a metamaterial. The two-port antenna design proposed uses a complementary split-ring resonator patch. The design results are also compared with a simple patch antenna to show the improvement. The design shows a better isolation of 50 dB. A broadband width of 8.3 THz is achieved using this complementary split-ring resonator design. The percentage bandwidth is 90%, showing an ultrabroadband response. The highest gain of 10.34 dB is achieved with this design. Structural parametric optimization is applied to the complementary split-ring resonator MIMO antenna design. The designed antenna is also optimized by applying parametric optimization to different geometrical parameters. The optimized design has a 20 μm ground plane, 14 μm outer ring width, 6 μm inner ring width, and 1.6 μm substrate thickness. The proposed antenna with its broadband width, high gain, and high isolation could be applied in high-speed communication devices.

Keywords: MIMO; metamaterial; antenna; optimization; THz; ultrabroadband; high gain; high isolation

Citation: Armghan, A.; Aliqab, K.; Alsharari, M.; Alsalman, O.; Parmar, J.; Patel, S.K. Design and Development of Ultrabroadband, High-Gain, and High-Isolation THz MIMO Antenna with a Complementary Split-Ring Resonator Metamaterial. *Micromachines* **2023**, *14*, 1328. https://doi.org/10.3390/mi14071328

Academic Editors: Trushit Upadhyaya, Hari Shankar Singh and Mark L. Adams

Received: 30 April 2023
Revised: 25 May 2023
Accepted: 10 June 2023
Published: 29 June 2023

Copyright: © 2023 by the authors. Licensee MDPI, Basel, Switzerland. This article is an open access article distributed under the terms and conditions of the Creative Commons Attribution (CC BY) license (https://creativecommons.org/licenses/by/4.0/).

1. Introduction

Antennas are a type of transducer used for communicating wirelessly between two devices. Antennas, which in the past were designed to have a huge aperture size, have now been reduced to small and compact nanoantennas. This reduction in the size of the antenna also extensively reduced its gain; therefore, there is now a need for high-gain compact antennas. There are various ways of improving the gain and bandwidth of an antenna, and one of these ways is to incorporate metamaterials. Today, high-speed communication needs THz antennas to be operated at high speeds. The THz antenna has in recent times been researched by many researchers to be used in high-speed wireless communication devices. The THz antenna offers a higher bandwidth, which could be used to transfer more data at high speeds. The need for high-gain and high-bandwidth THz antennas has increased. The gain and bandwidth of these THz antennas can be improved using metamaterials.

Metamaterials are artificial materials that can be used to improve many parameters of THz antennas [1]. Split-ring resonators or thin wires can be used to create these metamaterials [2]. Complementary split-ring resonators can also be applied to THz antennas to improve their bandwidth and gain [3]. Metamaterial structures can be used to improve the frequency of THz peaks. Simulation results showing a peak enhancement of 0.5 THz were achieved using metamaterial structures [4]. A graphene-material-based metamaterial design has previously been used to improve the gain of the THz antenna design [5]. The design of a microstrip antenna was also presented for THz applications. The antenna was designed with metamaterial loading, offering a metamaterial realization of the circular split-ring resonator. Traditional microstrip patch antennas are produced to be smaller and using more efficient metamaterials [6]. For THz uses, current research delves into the topic of designing adjustable MIMO antennas with superior isolation, both in terms of their laterality and orthogonality placement. It is possible to tune antennas by changing the chemical potential of the patches composed of trapezoidal graphene loaded onto metallic patch antennas. With the help of flawed ground structures, mutual coupling between the MIMO antennas can be reduced, even when they are placed laterally [7]. A triple-band MPA built on a polyimide substrate and integrated with a metamaterial has been presented as a potential solution for future healthcare applications in the terahertz band [8]. An antenna with a tunable resonance frequency via an external voltage was presented through the use of a graphene material. The graphene SRR was used as a metamaterial element in the design. By taking advantage of the fact that the chemical potential of graphene can be altered separately for the patch and the array, it is possible to optimize the material's properties. A lens could then be incorporated into the design to improve the radiation properties of the proposed layout [9]. MIMO antennas are used in high-speed wireless communication applications [10]. MIMO antennas are applicable in 5G applications [11] and millimeter-wave applications [12].

Today, THz MIMO antennas are used in high-speed communication devices because of their good isolation, high gain, and high bandwidth. MIMO array antennas have been designed with graphene materials, and high isolation has been achieved using a serpentine resonator. The antennas have been applied in THz communication [13]. The two-port MIMO antenna design was created for THz communication where the graphene-based design provides high isolation with simultaneous transmit and receive modes [14]. One other two-port MIMO communication channel design has been used for THz applications, where the light in the design is maintained between MIMO channels. To maximize the system's throughput and reliability, parallel channels created using appropriately spaced antenna elements have been used, inspired by the principles of diffraction-limited optics [15]. Single-element monopole antennas have been extended into four-port MIMO antennas, and designed with improved isolation for THz applications. The bandwidth was improved using the four-element MIMO antenna design [16]. The graphene-material-based patch antenna design is reconfigurable through the change of the chemical voltage of the graphene material. This reconfigurable MIMO antenna is applicable in THz applications [17]. MIMO THz antennas are also applicable in the design of quantum key distribution. The MIMO antenna design has been shown to have a more powerful distribution compared to the single-element antenna design. The MIMO antenna has one extra secret key component compared to the simple antenna design [18], which is the tapered square patch design used for THz communication. The design is fed through using a microstrip line with a partial ground plane. This design has been shown to be capable of achieving good results for application in high-speed THz communication devices [19]. The MIMO antenna also provides better performance when loaded with metamaterials, which can be realized in the form of different components, such as thin wires, split-ring resonators, complementary split-ring resonators, etc. The improved MIMO antenna is applicable in many GHz [20] and THz applications [21].

The MIMO antenna design has been investigated for use in 5G communication, and has also been presented for integration with portable devices [22]. The use of the broadband THz four-port MIMO antenna covering a broad spectrum of frequencies was presented in [23]. The isolation of the MIMO antenna is important, as all the radiating elements need to be isolated from each other. High isolation is, thus, important, and one such MIMO antenna design with high isolation was presented in [24]. The loading of metamaterial elements onto the MIMO antenna design allows for the tunability of the polarization, which is important when used for differently polarized antennas [7]. The size of the antenna is also important if desired to use in portable applications. One such compact antenna for 5G applications was presented in [25]. High-gain and wideband antennas are applicable in many applications, as they cover most frequency bands for different applications. One such MIMO antenna for achieving a high-gain and wideband response was presented in [26] for vehicular applications.

The optimization of the structural design is essential in achieving the optimized parameters. The different types of optimization that can be applied to these designs are nonlinear parametric optimization and linear parametric optimization [27]. The selection of the optimization algorithm is based on the behavior of the response, whether it be linear or nonlinear [28]. Nonlinear parametric optimization has previously been presented and used to achieve high absorption for the design of a solar absorber [29]. A similar approach could also be applied for the design of an efficient antenna with enhanced parameters, such as the bandwidth, gain, etc. Mutual coupling between patch antenna array elements can be reduced through several approaches. One of the approaches was discussed in detail using a ladder resonator, presented in [30].

The growing demand for high-speed wireless communication devices has led the way toward the design of an antenna that works at the THz frequency. As such, we propose a MIMO antenna design that shows a broadband and high isolation response at the THz frequency range. The proposed antenna is designed with a CSRR-loaded patch antenna. The antenna design comprising a simple microstrip patch is also designed with similar dimensions to show the improvement in the design. Nonlinear parametric optimization is applied to different parameters to achieve the optimal MIMO antenna design. The proposed antenna could be essential for use in high-speed wireless communication devices. We show the design, results, and analysis in the upcoming sections.

2. THz MIMO Antenna Design

The THz MIMO antenna design was first prepared with a simple square patch, complemented by split-ring resonator etching, and, finally, through preparing the design of the complementary split-ring resonator patch antenna. The design is presented in Figure 1, showing different views for a better understanding of the design. The blue color indicates the dielectric substrate, and the gray part is the metal patch and ground plane. The complementary square split-ring resonator patch was 41 × 41 μm^2 in size. The two elements of this patch were used to prepare the two-port MIMO antenna. The MIMO antenna was fed through with a matched microstrip line, as shown in Figure 1a. The two-port MIMO antenna was placed over a 122 μm substrate with two resonating elements 20 μm apart from one another. The MIMO antenna was backed with a defected ground structure, which was used to improve the performance of the antenna. The ground was defected through etching a section of the ground plane. The width of the ground plane was then reduced from 61 μm to 20 μm. The thickness of the substrate layer was optimized to 1.5 μm. The optimization results are given in the following results section. The complementary split-ring resonator width was also optimized. The optimized value of the CSRR rings was R1 = 14 μm and R2 = 6 μm.

The antenna parameter calculation mainly depended on the following equation, where the length and width are inversely proportional to the frequency, as shown in Equations (1)–(4):

$$W = \frac{C}{2fr}\sqrt{\frac{2}{\varepsilon r + 1}} \tag{1}$$

$$\varepsilon eff = \frac{\varepsilon r + 1}{2} + \frac{\varepsilon r - 1}{2}\left[1 + 12\frac{h}{w}\right]^{-0.5} \tag{2}$$

$$\frac{\Delta L}{h} = 0.412\frac{(\varepsilon eff + 0.3)\left(\frac{w}{h} + 0.264\right)}{(\varepsilon eff - 0.258)\left(\frac{w}{h} + 0.8\right)} \tag{3}$$

$$L = \frac{1}{2fr\sqrt{\varepsilon eff \mu_0 \varepsilon_0}} - 2\Delta \tag{4}$$

The envelope correlation coefficient (ECC) and diversity gain (DG) could be calculated as per Equations (5) and (6) [31].

$$ECC = \frac{|S_{11}^{*}S_{12} + S_{21}^{*}S_{22}|^{2}}{\left(1 - \left(|S_{11}|^{2} + |S_{21}|^{2}\right)\right)\left(1 - \left(|S_{22}|^{2} + |S_{12}|^{2}\right)\right)} \quad (5)$$

The improvement in the SNR of the multiple-element system over a one-element system is referred to as diversity gain (DG). The DG was calculated using the following Equation (6).

$$DG = 10\sqrt{1 - |ECC|^{2}} \quad (6)$$

The results of the ECC and DG are discussed further in the results and discussions section.

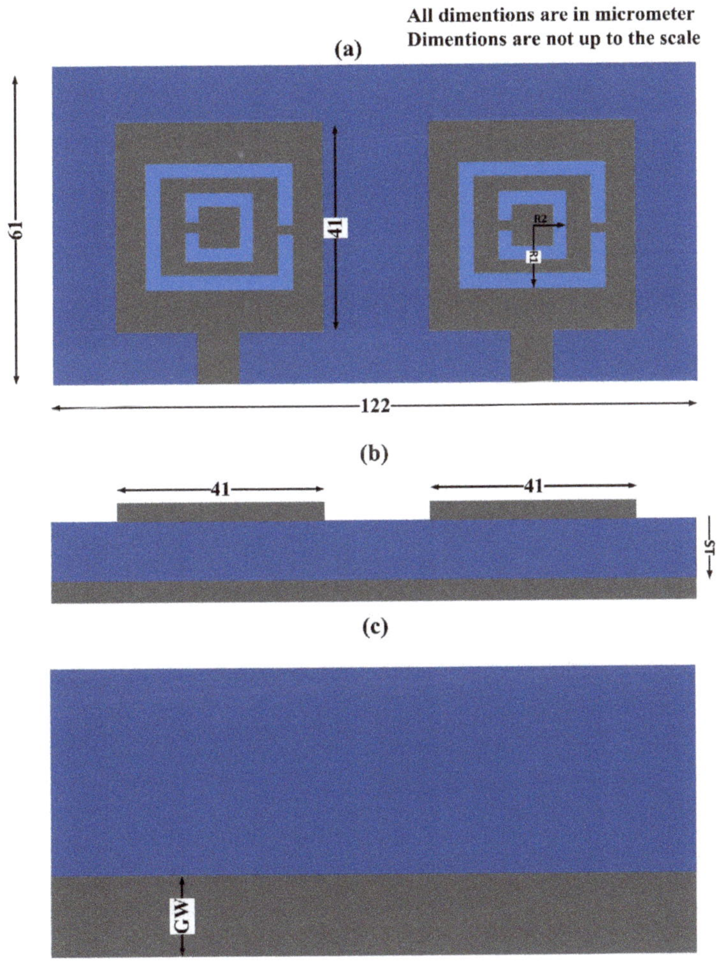

Figure 1. The CSRR metamaterial-based MIMO antenna design. (**a**) Top view of the THz MIMO antenna design showing the complementary split-ring resonator design. The patch was 41 × 41 μm². (**b**) Front view of the THz MIMO antenna design showing the MIM layer design. (**c**) Defected ground plane with 20 μm ground plane width (GW). The substrate and ground plane length was 122 μm. The width of the substrate was 61 μm. The substrate thickness (ST), inner ring width (R_2), and outer ring width (R_1) were 1.5 μm, 6 μm, and 14 μm, respectively.

3. THz MIMO Design Results

The design presented in Figure 1 was analyzed to obtain its results for different S-parameters. The square patch design 41 × 41 μm^2 in size was simulated first, and its response in terms of the S-parameters and gain was obtained. The square patch design was then etched with a split-ring resonator, and a complementary split-ring resonator metamaterial patch design was obtained, given in Figure 1. The new metamaterial design was analyzed and its results were obtained and shown in Figure 2. The results for the simple patch MIMO design were compared with the complementary split-ring resonator metamaterial patch antenna design, with the comparison showing that the metamaterial design had better performance in terms of the S-parameters and gain. The bandwidth obtained was also higher for the metamaterial design. The bandwidth obtained for the CSRR metamaterial patch MIMO antenna design was 8.3 THz, showing its ultrabroadband behavior. This high bandwidth could be used for the development of higher bandwidth high-speed communication system designs. The simple patch MIMO antenna design had the highest bandwidth of 1 THz between 14 THz and 15 THz. The simple MIMO antenna design gave four bands with the highest bandwidths of 1 THz. The highest isolation of approximately 50 dB was achieved for the CSRR metamaterial patch MIMO antenna design. The design's results were further optimized using changes in various physical parameters, such as substrate height, ground plane width, outer ring width, and inner ring width. The results for the different MIMO parameters, such as ECC and DG, were analyzed and presented in this section. The gain results for the two designs were also analyzed and presented in Figure 3. The antenna designed in this research could be easily fabricated through the use of lithography and by placing the metal patch over the substrate and etching the patch to create a CSRR shape.

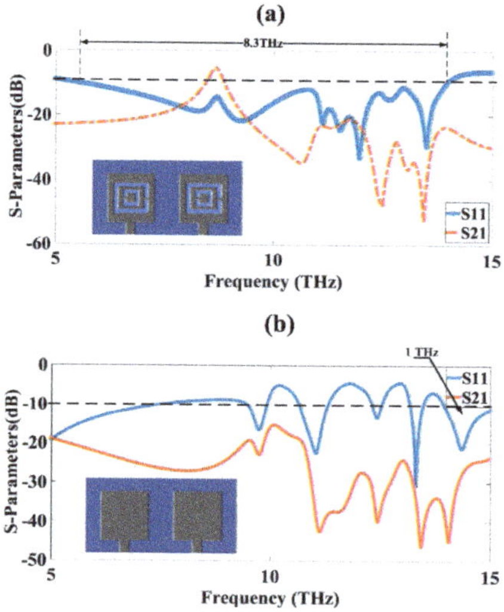

Figure 2. S-parameter results in dB for THz MIMO antenna designs. (**a**) CSRR metamaterial-loaded patch antenna. (**b**) Simple patch design. The patch was 41 × 41 μm^2 backed with 122 × 61 μm^2 SiO_2 substrate. The ground plane and patch were composed of a gold material. The thickness of the ground plane and patch was 500 nm.

Table 1. CSRR metamaterial MIMO antenna design comparison with simple patch MIMO antenna design.

Design	Bandwidth (THz)	Gain (dB)	Isolation (dB)
CSRR metamaterial MIMO antenna design	8.3	10.34	50
Simple patch MIMO antenna design	1	4.18	45

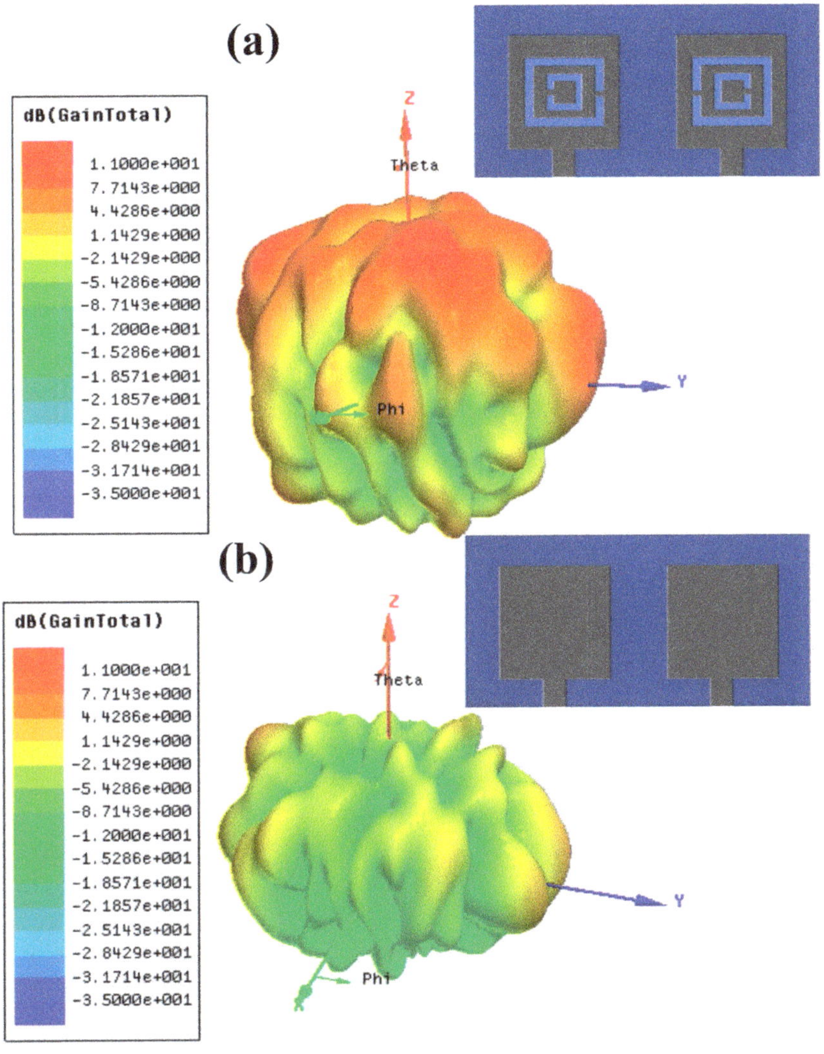

Figure 3. Gain results for both MIMO designs in dB. (**a**) CSRR metamaterial-loaded patch antenna. (**b**) Simple patch design. The highest gain of the CSRR metamaterial-loaded patch design was 10.34 dB, while the simple patch design was 4.18 dB. The metamaterial inclusion improved the gain by more than double its original value. The comparison of the two designs for different antenna parameters is presented in Table 1. The comparison showed that the CSRR metamaterial MIMO antenna design outperformed the simple patch MIMO antenna design.

The highest gain for the simple patch MIMO antenna design was 4.3 dB, as shown in Figure 3b. The results presented in the figure showed that the gain for the simple patch antenna design was on the lower side. The addition of metamaterial loading improved the gain of the design. The CSRR metamaterial patch MIMO antenna resulted in 10.34 dB, which was greater than the simple patch MIMO design. The increase in the gain was achieved through the etching of the metamaterial in the simple patch design. The etching of the split-ring resonator from the patch changed its permittivity and permeability, which resulted in the improved response of the antenna design.

3.1. MIMO Antenna Parameter Analysis

The different MIMO antenna parameters, such as the diversity gain (DG) and envelope correlation coefficient (ECC), were calculated in this section to show their effect on the analyzed spectrum ranging from 5 to 15 THz. Both parameters were analyzed using Equations (5) and (6). The S-parameter results were applied to the equation, and the results achieved for the ECC and DG were given in Figure 4. The ideal value for the DG was 10 dB, but because of losses, it was not achieved fully. The diversity gain reached a value of 10 dB throughout the studied spectrum from 5 to 15 THz, except in some parts of the spectrum at approximately 9 THz. The diversity gain value decayed for some frequencies at approximately 9 THz, which presented the idea that the diversity was inadequate only in this frequency range. The S-parameter curves and isolation between the two antennas was weak around that frequency, which showed that the diversity was also weak, reducing the diversity gain at that point. Similarly, the way the ECC increased at the same frequency also showed that the behavior of the design was adequate in the frequency range of 5 THz to 15 THz. The diversity was good overall for the data transmission and reception in the studied range. The ideal diversity value would have been zero, but there were still some minor values available.

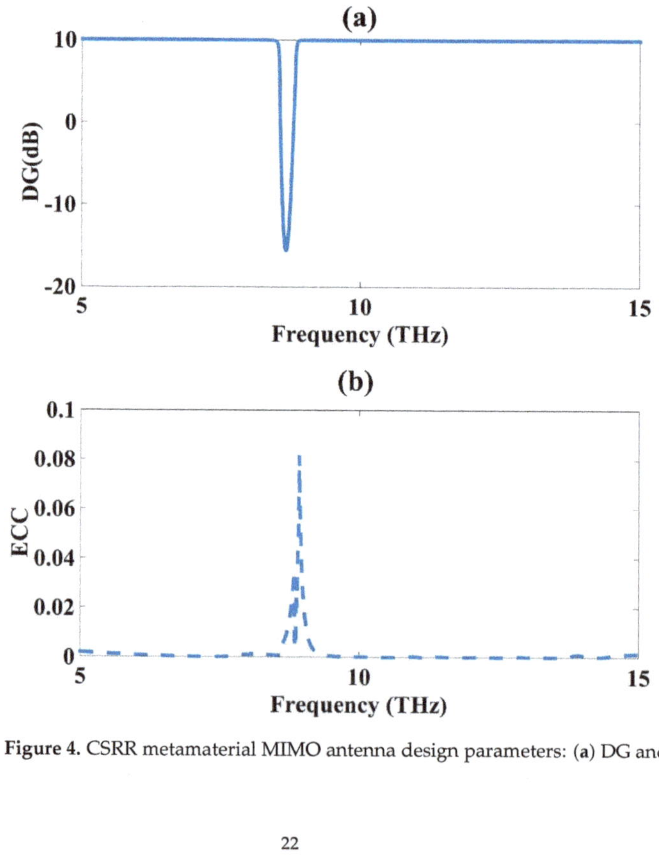

Figure 4. CSRR metamaterial MIMO antenna design parameters: (**a**) DG and (**b**) ECC.

3.2. Structural Parameter Optimization

The optimization algorithms could be applied to different structural parameters to obtain optimized parameters that give not only the best results, but also a compact design. The parametric optimization method can be applied to these structural parameters to obtain different optimized parameters. There are two main types of parametric optimization algorithms [32]. The first one is linear parametric optimization, and the second is the nonlinear parametric optimization method [33]. The selection of this method is based on the behavior of the design, whether the behavior of the design is linear or nonlinear. Based on this behavior, the correct type of optimization algorithm can be applied. The results of the reflectance clearly showed that the behavior of the results was nonlinear, and, thus, the nonlinear parametric optimization method could be applied to obtain the optimized structural parameters.

The functions did not behave linearly, which gave this optimization. It had function $f(x)$, constraint $c_i(x) = 1, 2, \ldots n$, or $d_j(x) = 1, 2, \ldots n$, which are components of x that were nonlinear [34].

The optimization of different structural parameters, such as the substrate height, ground plane width, and CSRR ring widths, was carried out to optimize antenna results, such as the bandwidth and S-parameters.

3.3. Substrate Thickness (ST) Optimization

The ST optimization was carried out to obtain the highest bandwidth and good results for the MIMO antenna design. The optimization was applied to the CSRR metamaterial-loaded MIMO antenna design. The substrate of the design varied from 0.5 μm to 1.5 μm to observe its effect on the absorption results. The variation was kept at 1.5 μm, because increasing the substrate more than this would increase the overall area of the structure, as well as the cost of the structure, so it was preferential to increase it to a certain limit, keeping it to a 1.5 μm thickness, so that the substrate was kept to a limit, suitable for fabrication and also reducing the cost of the substrate. The variation in the figure clearly showed that the reflectance shown in Figure 5a had the highest bandwidth for the 1.5 μm thickness. The reflectance level was also high for this value only. The yellow color curve showed the reflectance for the 1.5 μm thickness. The red and blue color curves showed a low bandwidth and low reflectance. The transmittance of the design is presented in Figure 5b, also showing that the transmittance results of the yellow curve of the 1.5 μm thickness were −50 dB higher, which showed that there was a 50 dB isolation, which is very good for MIMO antennas. The optimized value of the ST was 1.5 μm.

3.4. Ground Layer Width (GW) Optimization

The GW optimization was carried out to obtain the highest bandwidth and good results for the MIMO antenna design. The optimization was applied to the CSRR metamaterial-loaded MIMO antenna design. The ground layer width of the design varied from 20 μm to 35 μm to observe its effect on the absorption results. The variation was kept from 20 μm to 35 μm, because increasing the ground layer width further would have given abrupt results, and the defected ground concept was implemented by etching the part of the ground plane. The variation in the figure clearly showed that the reflectance shown in Figure 6a had the highest bandwidth for the 20 μm ground layer width. The S_{21} results of the design are presented in Figure 6b, showing that for 20 μm, the average value was less than −20 dB, with the highest depth obtained at approximately −50 dB. When the ground layer width was increased to 25, the reflectance showed only one band with a much lower bandwidth, and the transmittance also showed a low isolation compared to the 20 μm results. The increase in the ground layer width did not improve the results any further. Thus, the optimized value of the GW obtained was 20 μm.

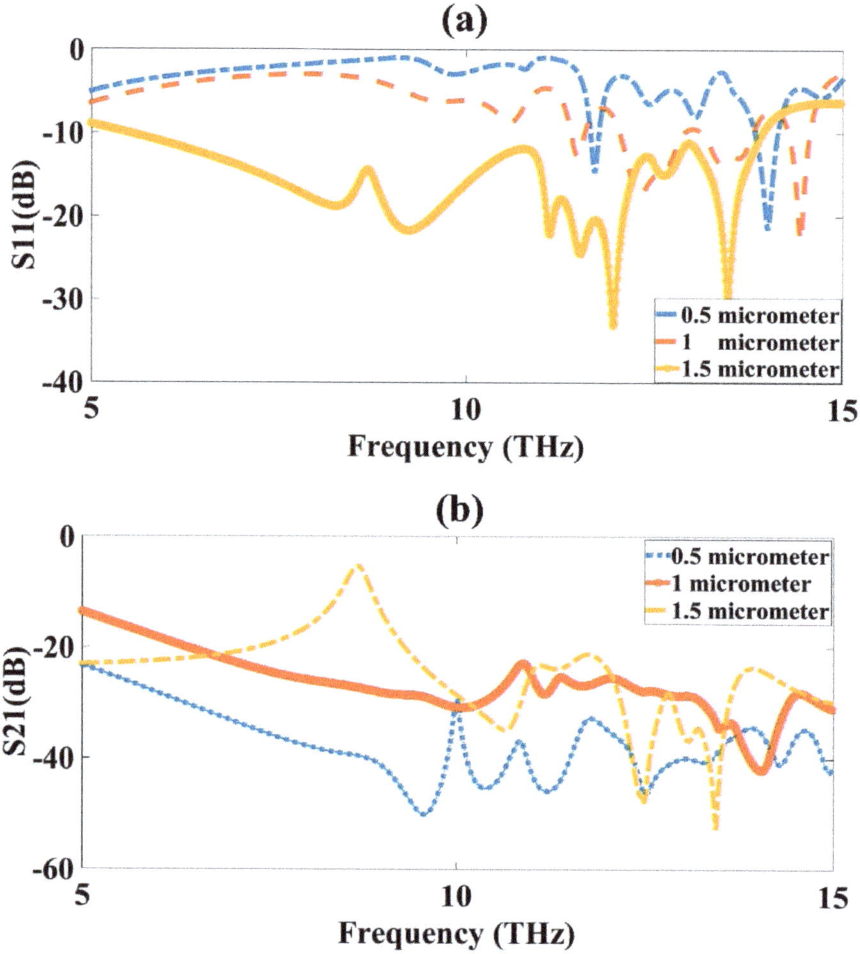

Figure 5. ST variation from 0.5 μm to 1.5 μm. (**a**) S_{11} and (**b**) S_{21}. The variation was carried out for a 5 to 15 THz frequency range. The optimized value for the ST was 1.5 μm, shown as the yellow-colored curve in the figure.

3.5. Inner Ring Width (R_2) Optimization

The inner ring width (R_2) optimization was carried out to obtain the highest bandwidth and good results for the MIMO antenna design. The optimization was applied to the CSRR metamaterial-loaded MIMO antenna design. The inner ring width of the design varied from 6 μm to 8 μm to observe its effect on the absorption results. The variation was kept at 6 μm to 8 μm, because increasing the width of the inner ring further would mix with the outer ring of the CSRR. The change in the ring width would change the capacitance of the metamaterial design. The increase in the width increased the capacitance and degraded the results, as shown in Figure 7. The blue-colored curve in Figure 7a showed the maximum bandwidth of 6 μm. The increase in the inner ring width degraded the response, and there was a reduction in the bandwidth, with more resonating bands available at a lower bandwidth. When the width increased to 8 μm, the result worsened, and there was a mismatch in the power that also resulted in one peak of 20 dB reflectance, which was not valid and, therefore, we could not consider this width for the design of the antenna.

The results showed a mismatch as we further increased the inner ring width; thus, only these three values were considered in this result. The S_{21} results also showed that for the increased inner width, the results of the transmittance showed a peak of 20 dB, which was not valid and could not be considered. The optimized value of the inner ring width was 6 μm.

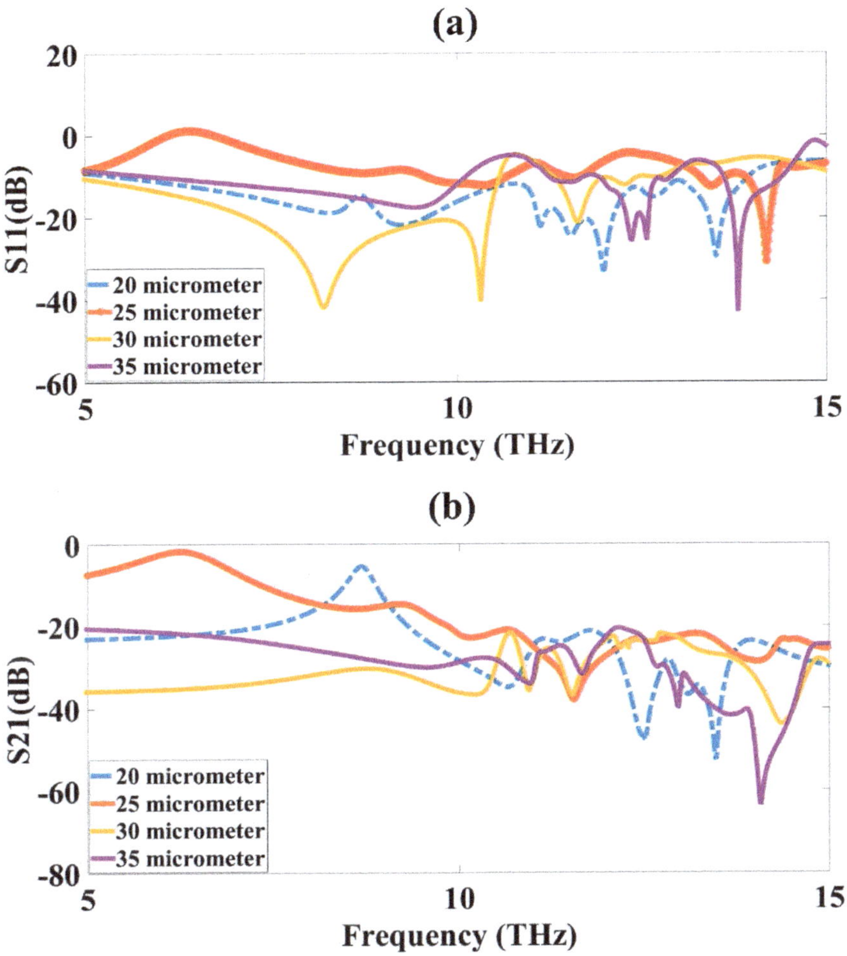

Figure 6. GW variation from 20 μm to 35 μm. (**a**) S_{11} and (**b**) S_{21}. The variation was carried out for the 5 to 15 THz frequency range. The optimized value of GW was 20 μm, shown as the blue-colored curve in the figure.

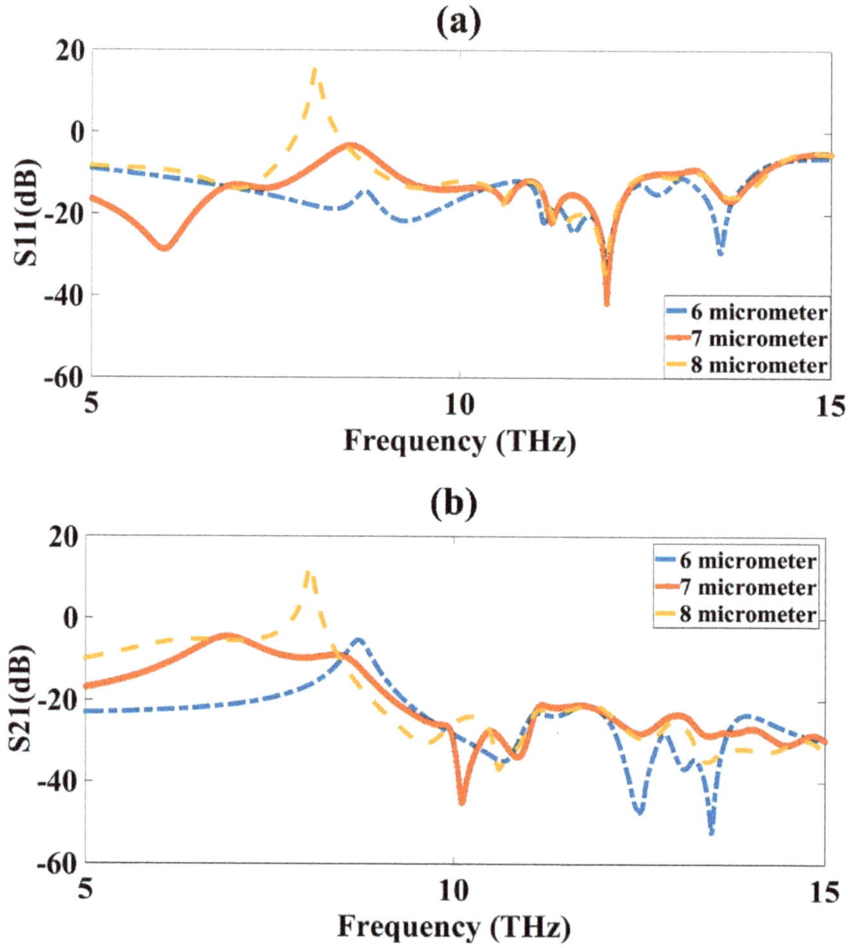

Figure 7. Inner ring width (R_2) variation from 6 μm to 8 μm. (**a**) S_{11} and (**b**) S_{21}. The variation was carried out for 5 to 15 THz frequency range. The optimized value of inner ring width (R_2) was 6 μm, shown as the blue-colored curve in the figure.

3.6. Outer Ring Width (R_1) Optimization

The outer ring width (R_1) optimization was carried out to obtain the highest bandwidth and good results for the MIMO antenna design. The optimization was applied to the CSRR metamaterial-loaded MIMO antenna design. The outer ring width of the design varied from 14 μm to 16 μm to observe its effect on the absorption results. The variation was kept at 14 μm to 16 μm, because increasing the width of the outer ring would increase the capacitance of the structure. The increase in the width increased the capacitance and degraded the results, as shown in Figure 8. The blue-colored curve in Figure 8a showed the maximum bandwidth of 14 μm. The reflectance for the 14 μm outer width thickness showed better results, giving the maximum bandwidth compared to all the other investigated design lengths. The 15 μm outer ring width was shown with an orange, dashed color plot in the figure, with the plot showing less than −10 dB results for three bands with a maximum bandwidth at approximately 1 THz, which was much lower compared to the 8.3 THz bandwidth of the 14 μm width design. The results were even more degrading for the 16 μm design results, which showed only one band with at approximately −12 dB reflectance, and with a lower bandwidth. Similar results were also obtained for S_{21}, with the results presented in Figure 8b. Thus, the optimized outer ring width obtained was 14 μm.

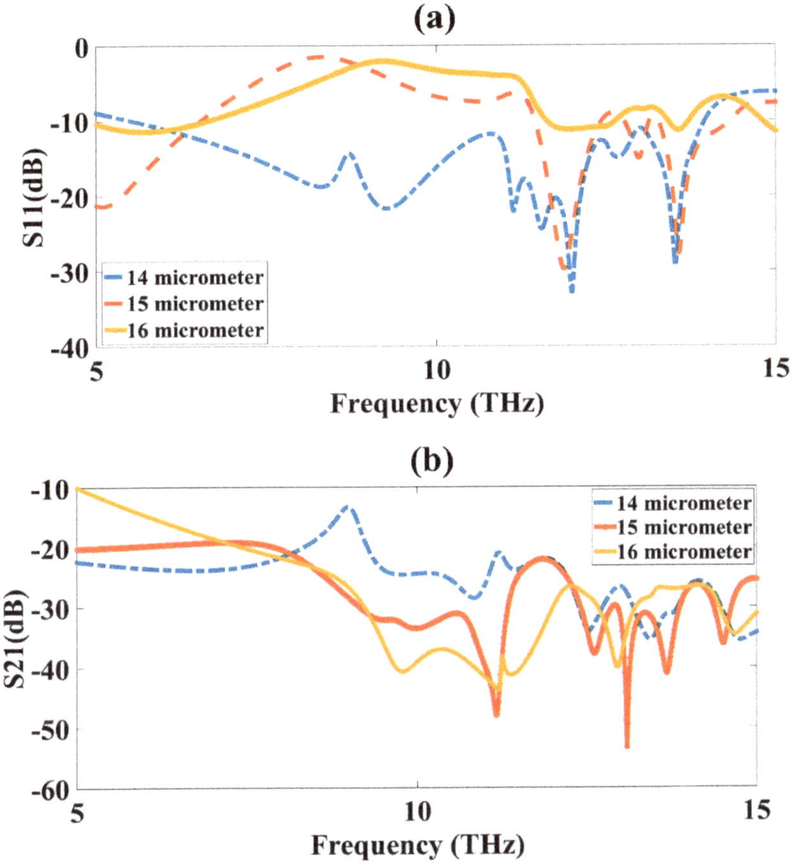

Figure 8. Outer ring width (R_1) variation from 14 μm to 16 μm. (**a**) S_{11} and (**b**) S_{21}. The variation was carried out for 5 to 15 THz frequency range. The optimized value of inner ring width (R_1) was 14 μm, shown as the blue-colored curve in the figure.

The permittivity, which is very important for metamaterials, was also obtained and presented in Figure 9. Both the real and imaginary parts were presented. The complementary split-ring resonator metamaterial was placed in a two-port network and the permittivity was calculated based on the metamaterial approach given in [35]. To be a metamaterial, its permittivity would have to be negative and in the figure; it was visible that the permittivity attained negative values in the investigated spectrum. The current distribution in the MIMO antenna design is also presented in Figure 10. The current distribution showed the highest current density of 3×10^4 A/m. The 2D radiation pattern plot is also presented in Figure 11 for different phi values. Four different radiation patterns were presented for phi values of 45°, 90°, 135°, and 180°. Radiation patterns were presented for a reference frequency of 10 THz. The radiation patterns for the other frequencies and other phi values could also be similarly achieved.

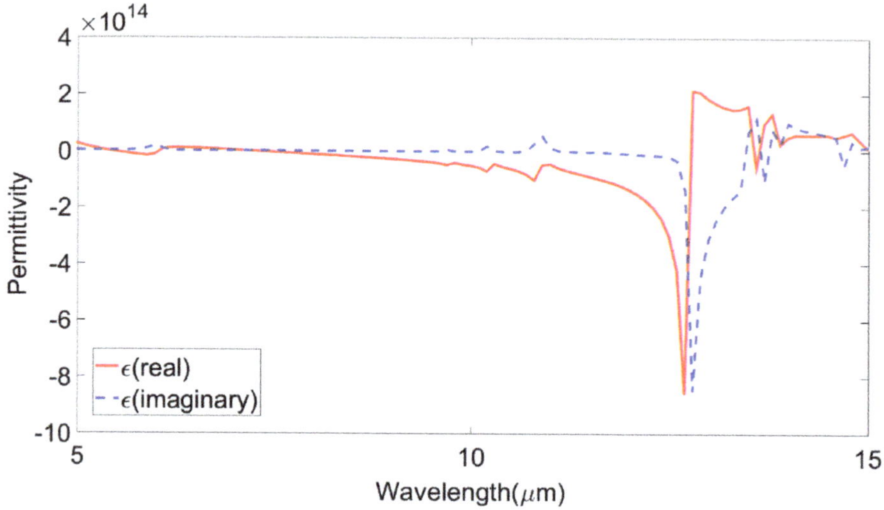

Figure 9. The permittivity curve for the investigated wavelength range of 5 μm to 15 μm. The real part and imaginary parts of permittivity are presented as red-colored and blue-colored dashed lines, respectively.

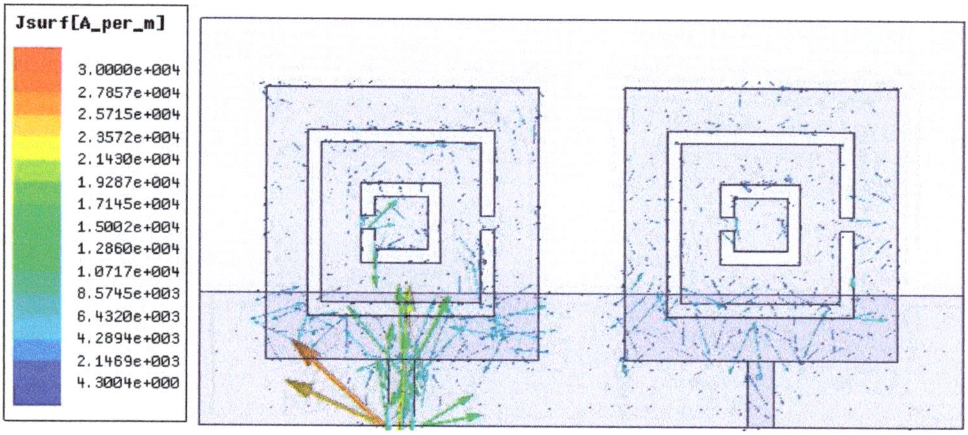

Figure 10. The current distribution in the MIMO antenna design: 3×10^4 A/m.

The THz MIMO antenna design for the CSRR metamaterial patch element and simple patch element was compared with outer designs and presented in Table 2. From the comparison, it was clear that the design had a high bandwidth and high gain. Apart from this, the design of the antenna was simple to fabricate and low cost, as we achieved everything with a two-port MIMO design only compared to a four-port design. Thus, the novelty of the design was shown to be a high bandwidth, high gain, low cost, easy fabrication, and compact size. These were the five novelties associated with our metamaterial MIMO design.

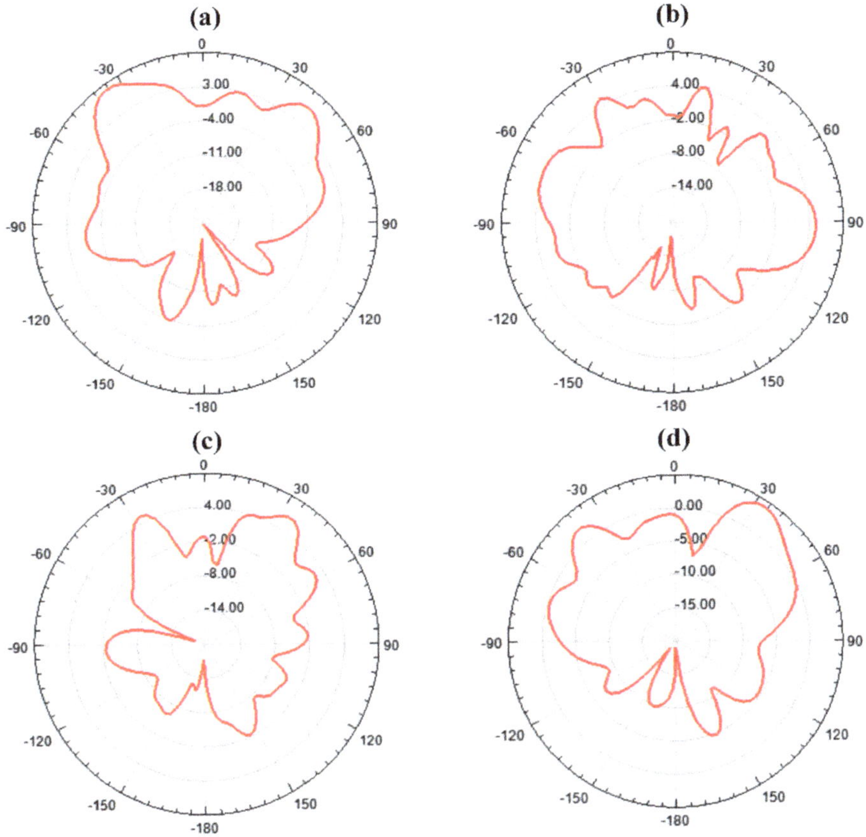

Figure 11. Radiation patterns for different theta values for 10 THz frequency. (**a**) Phi = 45°; (**b**) Phi = 90°; (**c**) Phi = 135°; (**d**) Phi = 180°.

Table 2. CSRR metamaterial MIMO antenna design comparison with simple patch MIMO antenna design and other designs.

Design	Bandwidth (THz)	Gain (dB)	Isolation (dB)
CSRR metamaterial MIMO antenna design	8.3	10.34	50
Simple patch MIMO antenna design	1	4.18	45
[7]	1.4	-	38
[36]	0.6	7.23	55
[37]	0.15	5	50
[15]	-	7.69	-
[38]	1.25	5.72	30
[39]	0.5	3.9	52
[40]	0.12	-	30
[41]	1	-	-

4. Conclusions

We compared two THz MIMO antenna designs in our research, namely, a CSRR metamaterial design and a square patch design. The gain and S-parameters were compared for these two designs. The CSRR design showed better performance compared to the other design at a 50 dB isolation, 10.34 dB gain, and 8.3 THz ultrabroadband width. The investigation was carried out for a 5 to 15 THz frequency range. Nonlinear parametric optimization was applied to the ground layer width, substrate thickness, CSRR inner ring width, and CSRR outer ring width. The optimized design parameters were achieved through this optimization. The current distribution and radiation patterns were also presented for the CSRR metamaterial design. The permittivity of the CSRR metamaterial design showed a negative behavior for its real and imaginary parts. The designed MIMO antennas were also compared with other published works. Overall, the proposed MIMO antenna with its high isolation, high gain, and broadband response could be applied in high-speed wireless communication devices.

Author Contributions: Conceptualization, S.K.P. and A.A.; methodology, S.K.P. and A.A.; software, all authors.; resources, K.A., O.A. and M.A.; writing—original draft preparation, all authors; writing—review and editing, all authors. All authors have read and agreed to the published version of the manuscript.

Funding: Researchers Supporting Project number (RSPD2023R654), King Saud University, Riyadh, Saudi Arabia.

Data Availability Statement: The data will be made available at a reasonable request to the corresponding author.

Conflicts of Interest: The authors declare no conflict of interest.

References

1. Yang, Y.J.; Wu, B.; Zhao, Y.T.; Fan, C. Dual-band beam steering THz antenna using active frequency selective surface based on graphene. *EPJ Appl. Metamaterials* **2021**, *8*, 12. [CrossRef]
2. Shelby, R.A.; Smith, D.R.; Schultz, S. Experimental verification of a negative index of refraction. *Science* **2001**, *292*, 77–79. [CrossRef]
3. Geetharamani, G.; Aathmanesan, T. Split ring resonator inspired THz antenna for breast cancer detection. *Opt. Laser Technol.* **2020**, *126*, 106111. [CrossRef]
4. Alizadeh, A.; Nazeri, M.; Sajedi Bidgoli, A. Enhancement of the frequency peak of terahertz photoconductive antennas using metamaterial (MTM) superstrate structures. *J. Comput. Electron.* **2020**, *19*, 451–456. [CrossRef]
5. Seyedsharbaty, M.M.; Sadeghzadeh, R.A. Antenna gain enhancement by using metamaterial radome at THz band with reconfigurable characteristics based on graphene load. *Opt. Quantum Electron.* **2017**, *49*, 221. [CrossRef]
6. Althuwayb, A.A. On-Chip Antenna Design Using the Concepts of Metamaterial and SIW Principles Applicable to Terahertz Integrated Circuits Operating over 0.6-0.622 THz. *Int. J. Antennas Propag.* **2020**, *2020*, 6653095. [CrossRef]
7. Das, P.; Singh, A.K.; Mandal, K. Metamaterial loaded highly isolated tunable polarisation diversity MIMO antennas for THz applications. *Opt. Quantum Electron.* **2022**, *54*, 250. [CrossRef]
8. Ashyap, A.Y.I.; Alamri, S.; Dahlan, S.H.; Abidin, Z.Z.; Inam Abbasi, M.; Majid, H.A.; Kamarudin, M.R.; Al-Gumaei, Y.A.; Hashim Dahri, M. Triple-Band Metamaterial Inspired Antenna for Future Terahertz (THz) Applications. *Comput. Mater. Contin.* **2022**, *72*, 1071–1087. [CrossRef]
9. Radwan, A.; D'Amico, M.; Din, J.; Gentili, G.G.; Verri, V. Bandwidth and gain enhancement of a graphene-based metamaterial antenna for the THz band. *ARPN J. Eng. Appl. Sci.* **2016**, *11*, 6349–6354.
10. Sharma, M.; Vashist, P.C.; Alsukayti, I.; Goyal, N.; Anand, D.; Mosavi, A.H. A wider impedance bandwidth dual filter symmetrical mimo antenna for high-speed wideband wireless applications. *Symmetry* **2022**, *14*, 29. [CrossRef]
11. Mohammed, S.L.; Alsharif, M.H.; Gharghan, S.K.; Khan, I.; Albreem, M. Robust hybrid beamforming scheme for millimeter-wave massive-MIMO 5G wireless networks. *Symmetry* **2019**, *11*, 1424. [CrossRef]
12. Alsaif, H.; Eleiwa, M.A.H. Compact design of 2 × 2 mimo antenna with super-wide bandwidth for millimeters wavelength systems. *Symmetry* **2021**, *13*, 233. [CrossRef]
13. Vijayalakshmi, K.; Selvi, C.S.K.; Sapna, B. Novel tri-band series fed microstrip antenna array for THz MIMO communications. *Opt. Quantum Electron.* **2021**, *53*, 395. [CrossRef]
14. Ali, M.F.; Bhattacharya, R.; Varshney, G. Graphene-based tunable terahertz self-diplexing/MIMO-STAR antenna with pattern diversity. *Nano Commun. Netw.* **2021**, *30*, 100378. [CrossRef]
15. Khalid, N.; Akan, O.B. Experimental Throughput Analysis of Low-THz MIMO Communication Channel in 5G Wireless Networks. *IEEE Wirel. Commun. Lett.* **2016**, *5*, 616–619. [CrossRef]

16. Raj, U.; Kumar Sharma, M.; Singh, V.; Javed, S.; Sharma, A. Easily extendable four port MIMO antenna with improved isolation and wide bandwidth for THz applications. *Optik* **2021**, *247*, 167910. [CrossRef]
17. Xu, Z.; Dong, X.; Bornemann, J. Design of a reconfigurable MIMO system for THz communications based on graphene antennas. *IEEE Trans. Terahertz Sci. Technol.* **2014**, *4*, 609–617. [CrossRef]
18. Kundu, N.K.; Dash, S.P.; McKay, M.R.; Mallik, R.K. MIMO Terahertz Quantum Key Distribution. *IEEE Commun. Lett.* **2021**, *25*, 3345–3349. [CrossRef]
19. Keshwala, U. Microstrip line fed sinusoidal tapered square shaped MIMO antenna for THz applications. *Optik* **2021**, *247*, 167905. [CrossRef]
20. Dhasarathan, V.; Nguyen, T.K.; Sharma, M.; Patel, S.K.; Mittal, S.K.; Pandian, M.T. Design, analysis and characterization of four port multiple-input-multiple-output UWB-X band antenna with band rejection ability for wireless network applications. *Wirel. Networks* **2020**, *26*, 4287–4302. [CrossRef]
21. Nie, S.; Akyildiz, I.F. Beamforming in Intelligent Environments based on Ultra-Massive MIMO Platforms in Millimeter Wave and Terahertz Bands. In Proceedings of the ICASSP, IEEE International Conference on Acoustics, Speech and Signal Processing, Barcelona, Spain, 4–8 May 2020.
22. Iffat Naqvi, S.; Hussain, N.; Iqbal, A.; Rahman, M.; Forsat, M.; Mirjavadi, S.S.; Amin, Y. Integrated LTE and Millimeter-Wave 5G MIMO Antenna System for 4G/5G Wireless Terminals. *Sensors* **2020**, *20*, 3926. [CrossRef] [PubMed]
23. Mohanty, A.; Sahu, S. A Micro 4-port THz MIMO antenna for nano communication networks. *Photonics Nanostructures-Fundam. Appl.* **2023**, *53*, 101092. [CrossRef]
24. Saxena, G.; Chintakindi, S.; Kasim, M.A.; Maduri, P.K.; Awasthi, Y.K.; Kumar, S.; Kansal, S.; Jain, R.; Sharma, M.K.; Dewan, C. Metasurface inspired wideband high isolation THz MIMO antenna for nano communication including 6G applications and liquid sensors. *Nano Commun. Netw.* **2022**, *34*, 100421. [CrossRef]
25. Sharma, M.K.; Sharma, A. Compact size easily extendable self isolated multi-port multi-band antenna for future 5G high band and sub-THz band applications. *Opt. Quantum Electron.* **2023**, *55*, 146. [CrossRef]
26. Prabhu, P.; Malarvizhi, S. Koch fractal loaded high gain Super-wideband diversity THz MIMO antenna for vehicular communication. *Opt. Quantum Electron.* **2022**, *54*, 726. [CrossRef]
27. Kler, A.M.; Zharkov, P.V.; Epishkin, N.O. Parametric optimization of supercritical power plants using gradient methods. *Energy* **2019**, *189*, 116230. [CrossRef]
28. Posypkin, M.; Khamisov, O. Automatic convexity deduction for efficient function's range bounding. *Mathematics* **2021**, *9*, 134. [CrossRef]
29. Alsaif, H.; Patel, S.K.; Ben Ali, N.; Armghan, A.; Aliqab, K. Numerical Simulation and Structure Optimization of Multilayer Metamaterial Plus-Shaped Solar Absorber Design Based on Graphene and SiO_2 Substrate for Renewable Energy Generation. *Mathematics* **2023**, *11*, 282. [CrossRef]
30. Roshani, S.; Shahveisi, H. Mutual Coupling Reduction in Microstrip Patch Antenna Arrays Using Simple Microstrip Resonator. *Wirel. Pers. Commun.* **2022**, *126*, 1665–1677. [CrossRef]
31. Kang, D.G.; Tak, J.; Choi, J. MIMO antenna with high isolation for WBAN applications. *Int. J. Antennas Propag.* **2015**, *2015*, 370763. [CrossRef]
32. Fang, H.; Fu, X.; Zeng, Z.; Zhong, K.; Liu, S. An Improved Arithmetic Optimization Algorithm and Its Application to Determine the Parameters of Support Vector Machine. *Mathematics* **2022**, *10*, 2875. [CrossRef]
33. Fang, Z.; Li, J.; Wang, X. Optimal control for electromagnetic cloaking metamaterial parameters design. *Comput. Math. Appl.* **2020**, *79*, 1165–1176. [CrossRef]
34. Aliqab, K.; Han, B.B.; Armghan, A.; Alsharari, M.; Surve, J.; Patel, S.K. Numerical Analysis and Structure Optimization of Concentric GST Ring Resonator Mounted over SiO_2 Substrate and Cr Ground Layer. *Mathematics* **2023**, *11*, 1257. [CrossRef]
35. Smith, D.R.; Vier, D.C.; Kroll, N.; Schultz, S. Direct calculation of permeability and permittivity for a left-handed metamaterial. *Appl. Phys. Lett.* **2000**, *77*, 2246–2248. [CrossRef]
36. Khaleel, S.A.; Hamad, E.K.I.; Parchin, N.O.; Saleh, M.B. MTM-Inspired Graphene-Based THz MIMO Antenna Configurations Using Characteristic Mode Analysis for 6G/IoT Applications. *Electronics* **2022**, *11*, 2152. [CrossRef]
37. Ali, M.F.; Bhattacharya, R.; Varshney, G. Tunable four-port MIMO/self-multiplexing THz graphene patch antenna with high isolation. *Opt. Quantum Electron.* **2022**, *54*, 822. [CrossRef]
38. Davoudabadifarahani, H.; Ghalamkari, B. High efficiency miniaturized microstrip patch antenna for wideband terahertz communications applications. *Optik* **2019**, *194*, 163118. [CrossRef]
39. Vasu Babu, K.; Das, S.; Varshney, G.; Sree, G.N.J.; Madhav, B.T.P. A micro-scaled graphene-based tree-shaped wideband printed MIMO antenna for terahertz applications. *J. Comput. Electron.* **2022**, *21*, 289–303. [CrossRef]
40. Ali, M.F.; Singh, R.K.; Bhattacharya, R. Re-configurable graphene-based two port dual-band and MIMO antenna for THz applications. In Proceedings of the 2020 IEEE Students' Conference on Engineering and Systems (SCES), Prayagraj, India, 10–12 July 2020.
41. Trichopoulos, G.C.; Mosbacker, H.L.; Burdette, D.; Sertel, K. A broadband focal plane array camera for real-time thz imaging applications. *IEEE Trans. Antennas Propag.* **2013**, *61*, 1733–1740. [CrossRef]

Disclaimer/Publisher's Note: The statements, opinions and data contained in all publications are solely those of the individual author(s) and contributor(s) and not of MDPI and/or the editor(s). MDPI and/or the editor(s) disclaim responsibility for any injury to people or property resulting from any ideas, methods, instructions or products referred to in the content.

micromachines

MDPI

Communication

Low-Profile Dual-Polarized Double-Layer Microstrip Antenna for 5G and 5G Wi-Fi

Wenxing An [1,2], Xiaoqing Tian [1], Jian Wang [1,2,*] and Shuangshuang Wang [3,4]

1 Tianjin Key Laboratory of Imaging and Sensing Microelectronic Technology, School of Microelectronics, Tianjin University, Tianjin 300072, China; anwenxing@126.com (W.A.); 2020232173@tju.edu.cn (X.T.)
2 Qingdao Key Laboratory of Marine Information Perception and Transmission, Qingdao Institute of Ocean Engineering, Tianjin University, Qingdao 266200, China
3 Tianjin Engineering Center of Integrated Circuit and Computing Systems, Tianjin 300072, China; 18526459322@163.com
4 Tianjin International Joint Research Center of Internet of Things, Tianjin 300072, China
* Correspondence: wangjian16@tju.edu.cn

Abstract: A dual-polarized double-layer microstrip antenna with a metasurface structure is proposed for 5G and 5G Wi-Fi. A total of 4 modified patches are used for the middle layer structure, and 24 square patches are used for the top layer structure. The double-layer design has achieved −10 dB bandwidths of 64.1% (3.13 GHz~6.08 GHz) and 61.1% (3.18 GHz~5.98 GHz). The dual aperture coupling method is adopted, and the measured port isolation is more than 31 dB. A low profile of $0.096\lambda_0$ is obtained (λ_0 is the wavelength of 4.58 GHz in the air) for a compact design. Broadside radiation patterns have been realized, and the measured peak gains are 11.1 dBi and 11.3 dBi for two polarizations. The antenna structure and E-field distributions are discussed to clarify its working principle. This dual-polarized double-layer antenna can accommodate 5G and 5G Wi-Fi simultaneously, which can be a competitive candidate for 5G communication systems.

Keywords: microstrip antenna; double-layer; low-profile; wideband; periodical structure; dual-polarized

Citation: An, W.; Tian, X.; Wang, J.; Wang, S. Low-Profile Dual-Polarized Double-Layer Microstrip Antenna for 5G and 5G Wi-Fi. *Micromachines* **2023**, *14*, 942. https://doi.org/10.3390/mi14050942

Academic Editor: Aiqun Liu

Received: 10 April 2023
Revised: 24 April 2023
Accepted: 25 April 2023
Published: 26 April 2023

Copyright: © 2023 by the authors. Licensee MDPI, Basel, Switzerland. This article is an open access article distributed under the terms and conditions of the Creative Commons Attribution (CC BY) license (https://creativecommons.org/licenses/by/4.0/).

1. Introduction

The popularity of intelligent wireless terminals, autopilot, Virtual Reality (VR), and the Internet of Things (IoT) has put forward high requirements on wireless communication networks in terms of large system capacity, high transmission speed, and low network delay. Compared with 4th Generation (4G) wireless communication technology, 5th Generation (5G) technology can achieve a high-speed data rate with reduced network delay. It can carry out stable transmission in different scenarios. Recently, 5G wireless systems have been deployed widely. In addition, 5G WiFi has been developed for wireless environments inside buildings to improve indoor wireless network coverage. It has strong anti-interference ability, broad bandwidth, high throughput, and scalability. Most 5G systems operate between 3.3 GHz and 5 GHz, and the 5G WiFi band is from 5.15 GHz to 5.85 GHz.

Antennas for 5G and 5G Wireless-Fidelity (Wi-Fi) can be applied in the fields of mobile communication for high-speed data transmission and low-latency communication services, vehicle networking to achieve safer and more intelligent transportation systems, industrial automation for large-scale IoT device connections and data transmission, and medical and health for real-time interaction between medical devices and cloud data.

Single-polarized antennas have been studied extensively [1–8]. However, polarization diversity antennas have several advantages over traditional single-polarized antennas. Firstly, they can provide better signal quality and reliability by reducing signal fading and interference caused by polarization mismatch, which is particularly important in urban or crowded environments with multiple signals. Secondly, polarization diversity antennas can improve the range of wireless communication systems by reducing the effects of multipath propagation. Multipath propagation occurs when signals are reflected off surfaces and arrive at the receiving antenna at different times and phases, causing distortion and interference. Using two antennas with different polarization orientations, the receiver can combine the signals from both antennas and effectively cancel out the interference caused by multipath propagation. Finally, polarization diversity antennas can increase the capacity of wireless communication systems by allowing multiple users to transmit and receive data simultaneously on the same frequency band without interfering with each other. Overall, polarization diversity antennas offer significant improvements in signal quality, reliability, range, and capacity compared to traditional single-polarization antennas, making them a popular choice for a wide range of wireless communication applications.

Many antennas with dual-polarization performance have been investigated widely, for example, dipole [9–14], patch [15–20], and slot designs [21–23]. Many dual-polarized antennas working at the 5G frequency band have been designed. For dipole designs, a pair of printed bent dipoles was introduced with a balun and parallel transmission line for wideband performance [9]. A bandwidth of 10.5% was obtained with a port isolation of 52 dB and cross-polarization of 40 dB. A dual-polarized fan-shaped dipole antenna was designed with a frequency range from 3.3 GHz to 4.2 GHz [10]. A dual-polarized dipole antenna was exploited in [11]; the metasurface structure was placed between the dipole and the ground for a low profile. A bandwidth greater than 29.5% was achieved with an isolation of 34 dB and a height of $0.096\lambda_0$. The bandwidth was extended by a metal plating antenna with a 3D printing process [12]. It consisted of two orthogonal polarized dipoles. The tested bandwidth was from 3.5 GHz to 5.1 GHz with the voltage standing wave ratios less than 1.5. The antenna height was $0.24\lambda_0$ and the isolation was better than 18 dB. A novel dual-polarized magneto-electric dipole was reported in [13] with dielectric substrate loading. The antenna structure included vertical short-circuit patches and horizontal planar dipoles. Based on this design, the antenna had an impedance matching bandwidth of 24.9% and a height of $0.15\lambda_0$. The bandwidth of the dual-polarized magneto-electric dipole antenna was further improved to 65.9% [14]. It had stable radiation patterns with a port isolation greater than 36 dB. However, the antenna had a relatively larger height of $0.24\lambda_0$.

For the patch designs, a single-ended, dual-polarized patch antenna was proposed in [15]. Although an extremely low profile of $0.06\lambda_0$ was realized, it had a limited bandwidth of 5.7% with a port isolation of 25.4 dB. A patch antenna was discussed in [16] with a hybrid feed structure. The -10 dB bandwidth was 14% and the isolation level between two ports was less than 40 dB. To expand the bandwidth, a differentially driven dual-polarized patch antenna was investigated [17]. A wide impedance matching frequency band of 17.2% was obtained from 3.17 GHz to 3.77 GHz with a low profile of $0.067\lambda_0$ (λ_0 is the free-space wavelength at 3.5 GHz). It had a high isolation greater than 38.5 dB and a low cross-polarization level less than -33 dB. A dual-polarized patch antenna was exploited with an etched bowtie slot [18]. A 10 dB reflection coefficient bandwidth of 18.8% was realized for two polarizations with a port isolation less than 28.5 dB and a low profile of $0.08\lambda_0$. To further improve the working frequency band, microstrip antenna designs using stacked patches have emerged [19,20]. A two-layer stacked patch was utilized to optimize the bandwidth [19]. A fractional impedance bandwidth of 19% was realized with a port isolation of 35 dB. In [20], a differentially fed stacked patch antenna was proposed for base station application. The driven patch was excited by tuned slots, and the top parasitic patch was employed to enhance the impedance bandwidth. A broadband performance of 49.4% was achieved with a high isolation greater than 37 dB.

For the slot designs, a differentially fed dual-polarized slot antenna was proposed for base station application [21]. Two H-shaped slots were etched on an octagon patch. It had a bandwidth of 19.3% with a VSWR (voltage standing wave ratio) less than 1.5 and a port isolation larger than 43 dB. The frequency band was improved by a microstrip-fed stepped-impedance slot antenna [22]. The slot was excited by a stepped feeding strip. The tested relative bandwidth was 38.7% from 1.69 GHz to 2.5 GHz with an isolation greater than 35 dB. The frequency band was further widened by a dual-polarized cross-shaped slot antenna, which was excited by U-shaped microstrip lines [23]. A broad bandwidth of up to 68% was achieved with an $|S_{11}|$ less than -10 dB from 1.3 GHz to 2.65 GHz. However, the port isolation was only 20 dB, and the profile was $0.27\lambda_0$.

There are also some other dual-polarized designs. A folded bowtie antenna was discussed in [24] operating between 3.1 and 5 GHz with a relatively large profile. The filtering antenna had a wide working band of 25.6% and a low cross-polarization ratio of 22 dB [25]. A differentially fed filtering antenna was reported for the 5G frequency band. It had dual operation bands from 3.28 GHz to 3.71 GHz and from 4.8 GHz to 5.18 GHz [26]. A wideband differentially fed laminated resonator antenna was reported in [27] with a wideband of 29% and a port isolation better than 35 dB.

In 2014, a broadband low-profile metasurface antenna was presented based on a periodical mushroom structure [28]. The proposed antenna was formed using 4×4 mushroom cells and a ground plane. The operating principle of the metasurface antenna was investigated, and the field distributions of the resonance modes were examined. Then, the metasurface antennas were studied extensively due to their desired features such as low-profile structure, stable radiation performance, high efficiency, and good cross-polarizations. Recently, metasurface antennas have been designed to accommodate different communication systems including 5G and 5G WiFi [29–32]. The bandwidth was extended to 28% in [29]. A dual-polarized metasurface antenna was reported in [30]. It had a low profile of $0.058\lambda_0$ with a -10 dB bandwidth of 25%, and the port isolation was greater than 34 dB. To improve the frequency band, a dual-polarized grid-slotted microstrip antenna was designed using a Y-shaped feeding strip [31]. The tested port isolation was better than 14.5 dB with a wideband performance of 43%. However, only part of the 5G band was covered. Then, a dual-polarized antenna with a low profile was designed for the 5G frequency band [32]. Its working band was from 3.2 GHz to 5.1 GHz with a bandwidth of 53.4%. However, only part of the 5G and 5G Wi-Fi bands were covered. So, some novel designs are expected to accommodate the 5G sub-6 GHz (3.3 to 5 GHz) and 5G Wi-Fi (5.15 GHz to 5.85 GHz) bands with a relative bandwidth of at least 55%. Furthermore, a low-profile design is more appreciated as it can save installation space effectively.

A comparison is conducted in Table 1 including different kinds of dual-polarized works. References [9,10,12,20,24] have relatively large profiles. References [13,21,22,25–27] have relatively narrow working frequency bands. Although references [11,15–19,29–32] have low-profile designs, their working frequency bands are relatively narrow. References [14,23] have broadband performances. However, the reported antenna heights were higher than $0.23\lambda_0$.

To achieve the required bandwidth, low profile, high isolation, and higher gain performance, we designed a dual-polarized low-profile broadband antenna to accommodate 5G and 5G Wi-Fi bands. The paper is organized as follows. The background of the dual-polarized antenna is discussed in Section 1. The detailed double-layer structure and materials of the proposed antenna elements are described in Section 2. To validate the dual-polarized design, a prototype with the metasurface was fabricated and tested in Section 3. The working mechanism, the influence of the double-layer structure, and the influence of the feeding strip on the antenna performance are investigated in Section 4. Some concluding remarks are given in Section 5.

Table 1. Comparison of this work with other dual-polarized designs.

Ref.	Height (λ_0 @ Center Freq.)	Relative Bandwidth (%)	Port Isolation (dB)	Element Peak Gain (dBi)	Ant. Type
[9]	$0.25\lambda_0$	10.5	52	9.9	dipole
[10]	$0.26\lambda_0$	24	25	7	dipole
[11]	$0.096\lambda_0$	29.5	20	9	dipole
[12]	$0.24\lambda_0$	37.2	18	9	dipole
[13]	$0.15\lambda_0$	24.9	29	8.2	dipole
[14]	$0.23\lambda_0$	65.9	36	9.5	dipole
[15]	$0.06\lambda_0$	5.71%	24	10.9	patch
[16]	$0.096\lambda_0$	14	40	7.4	patch
[17]	$0.067\lambda_0$	17.2	38.5	8.2	patch
[18]	$0.08\lambda_0$	18.8	28.5	8	patch
[19]	$0.08\lambda_0$	19	35	11	patch
[20]	$0.27\lambda_0$	49.4	37	8.7	patch
[21]	$0.13\lambda_0$	19.3	43	8.1	slot
[22]	$0.16\lambda_0$	38.7	35	N.A.	slot
[23]	$0.27\lambda_0$	68	20	9	slot
[24]	$0.45\lambda_0$	46.9	28	7.5	folded bowtie
[25]	$0.14\lambda_0$	25	20	N.A.	filtering ant.
[26]	$0.13\lambda_0$	12.3 and 7.6	37	8.34	filtering ant.
[27]	$0.115\lambda_0$	29.2	35	7.58	laminated ant.
[30]	$0.058\lambda_0$	25	34	10	metasurface
[31]	$0.06\lambda_0$	43	14.5	10.1	metasurface
[32]	$0.047\lambda_0$	53.4	30	9.9	metasurface

2. Materials and Methods

The proposed dual-polarized antenna is shown in Figure 1. The dual-polarized microstrip antennas consist of a double-layer metallic structure, a ground plane, and feeding strips printed on dielectric substrates. The double-layer patches are made of copper, and the substrates are typically made of insulating material. The ground plane serves as a reflector.

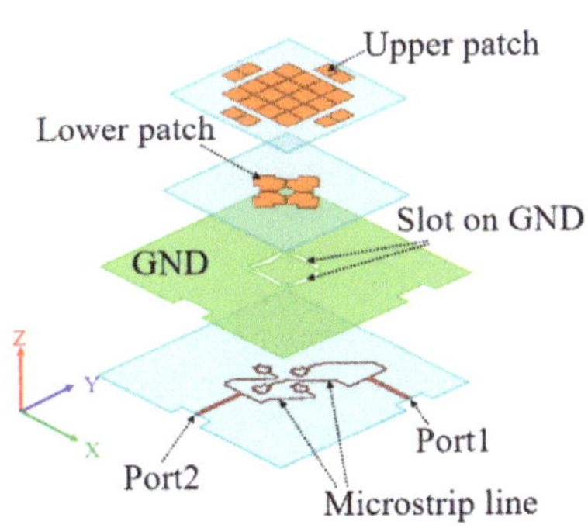

Figure 1. The proposed wideband dual-polarized double-layer antenna.

Relong and Epoxy glass fiber (FR4) laminates are employed for the top and middle layer substrates. The double-layer structure of the microstrip antennas has several advantages. It provides improved bandwidth and radiation patterns as the two layers can be designed to have different dielectric constants and thicknesses to achieve desired characteristics. The Relong and FR4 laminates have relative dielectric constants of 2.2 and 4.4, respectively. Their loss tangents are 0.0009 and 0.02. An air gap is used to separate the Relong and FR4 substrates. The use of low dielectric constant materials in microstrip antennas offers several advantages. Firstly, it can decrease the Q factor and improve the radiation efficiency of the antenna by reducing losses due to dielectric absorption. Secondly, it can increase the bandwidth of the antenna by reducing the effect of surface waves. So, low dielectric constant materials in microstrip antennas can improve the antenna's performance. Detailed antenna structural parameters are in Table 2.

Table 2. Antenna parameters (mm).

Para.	Value	Para.	Value	Para.	Value	Para.	Value	Para.	Value	Para.	Value	Para.	Value
L_G	125.0	L_6	22.0	L_{12}	2.0	L_{18}	2.5	W_1	10.5	W_7	2.4	G_1	1.1
L_1	31.0	L_7	24.2	L_{13}	5.0	L_{19}	4.0	W_2	8.7	W_8	3.0	G_3	4.6
L_2	28.2	L_8	9.0	L_{14}	3.8	L_{20}	7.5	W_3	15.0	W_9	19.5	R_1	4.0
L_3	11.8	L_9	20.1	L_{15}	3.4	L_{21}	7.5	W_4	5.0	W_{10}	18.0		
L_4	12.2	L_{10}	7.1	L_{16}	6.0	L_{22}	80.0	W_5	3.0	W_{11}	5.0		
L_5	4.8	L_{11}	10.0	L_{17}	8.5	L_{23}	83.0	W_6	2.3	G_1	0.8		

The double-layer structure is at the center of the ground with a size of 125×125 mm^2. The top layer has a height of 5.3 mm, and the adopted Relong substrate has a size of 80×80 mm^2 with a thickness of 1 mm. In Figure 2a, 16 square patches are printed at the center while 8 square patches are at the four edges for better impedance matching, especially at the higher band. The central patches have a separation of G_1 with a length of W_1. The edge patch has a length of W_2 with a gap of G_2. The middle layer is placed above the ground with a height of 2 mm. In Figure 2b, there are four metallic patches at the center of the 0.8 mm thick FR4 substrate with a size of 83×83 mm^2. The patch shape is optimized with a circular slot at the center.

In Figure 2c, the feed networks are placed at the bottom of the 31 mil thick Relong substrate. An aperture-coupled feeding method is adopted for the antenna excitation, which can decrease the current discontinuity on the patches. An aperture-coupled feeding technique has several advantages over other feeding methods. Firstly, aperture-coupled feeding provides a simple and compact structure for easy integration with other circuits and components. The feeding structure consists of a microstrip line with a slot coupled to the radiating element. The microstrip line can connect other circuits or components, such as filters or amplifiers, which makes it a versatile and flexible feeding method for various applications. Secondly, the coupling between the microstrip line and the radiating element can be adjusted by varying the gap size, providing more flexibility for a wider operating bandwidth compared to other feeding methods, such as coaxial or waveguide feeding. Overall, aperture-coupled feeding can offer a simple, compact, and high-performance feeding method for antenna design, making it a popular choice in many applications.

For each polarization in this design, two arc-shaped slots are carved on the ground, stimulated by the feedline on the bottom of this substrate. The feeding strip intersection would result in low port isolation [31]. A power divider is introduced into the feed structure, and the microstrip feedlines for two polarization directions are properly designed to avoid the intersection [30,32]. Good port isolation is achieved across the passband. The terminals of the feedline structure are modified for better impedance matching, which will be discussed later.

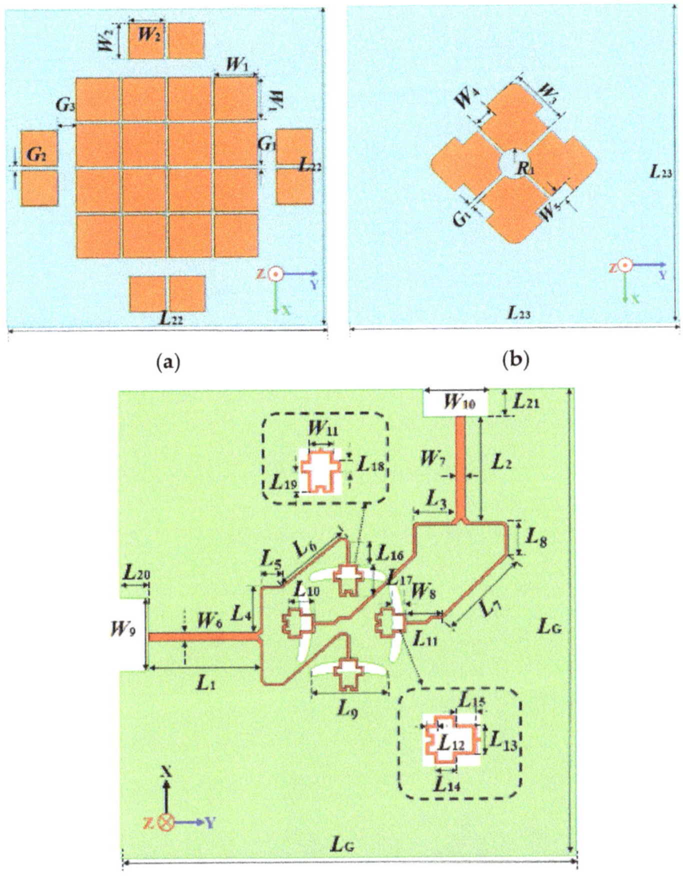

Figure 2. The dual-polarized antenna structure: (**a**) top structure; (**b**) middle layer; (**c**) ground plane.

3. Results

The double-layer antenna prototype was fabricated and assembled to verify the proposed antenna design. HFSS (High-Frequency Structure Simulator) software was adopted for antenna simulation and optimization. HFSS is a powerful 3D electromagnetic simulation software widely used in the design and analysis of high-frequency electronic components and systems. It uses the finite element method (FEM) to analyze the electromagnetic characteristics of three-dimensional objects. An air box was used to surround the antenna model. By setting radiation boundary conditions on the surface of an air box, an infinite space can be simulated. HFSS software uses adaptive mesh generation technology to automatically generate accurate and effective meshes to complete the discretization of analysis objects. Usually, the grid size is less than one-tenth of the wavelength corresponding to the solution frequency.

The antenna prototype is in Figure 3. The fabrication procedure was as follows: Firstly, the multilayer antenna structure was fabricated using printed circuit board technology. Then, the SMA connector was welded to the feed port on the antenna ground. Finally, the multilayer antenna structures were assembled, and the middle and upper structures were fixed on the ground at specific heights using plastic screws.

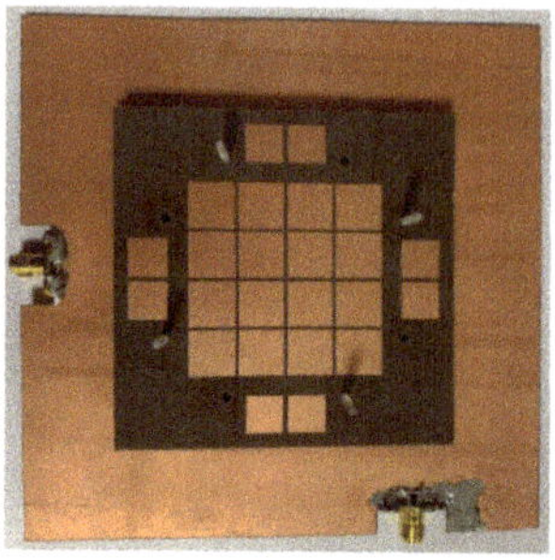

Figure 3. Antenna prototype.

A Rohde & Schwarz ZVA24 network analyzer was used for the measurement of the S-parameter, as shown in Figure 4. The parameters of the Rohde & Schwarz ZVA24 network analyzer are in Table 3. Two ports of the dual-polarized antenna were connected to the network analyzer, and the tested S-parameter results were measured and exported. Figure 5 depicts the calculated and tested results of the $|S_{11}|$, $|S_{22}|$, $|S_{12}|$, and gain. Although there were some assembly errors, the test and simulation results are consistent. For Port 1, the calculated frequency band less than -10 dB started from 3.29 GHz, and the tested band was from 3.13 GHz to 6.08 GHz with a relative bandwidth of 64.1%. For Port 2, the calculated band started from 3.3 GHz, and the measured band was from 3.18 GHz to 5.98 GHz with a relative bandwidth of 61.1%. Based on the tested results, the frequency band between 3.3 GHz and 5.875 GHz was completely covered.

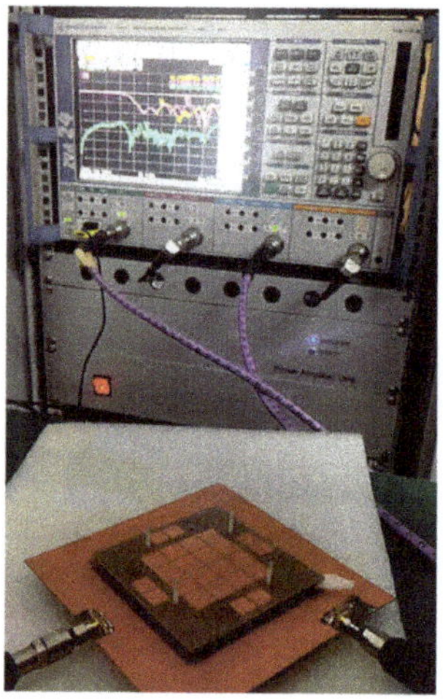

Figure 4. Antenna prototype testing scenario with Rohde & Schwarz ZVA24 network analyzer.

Table 3. The performance parameters of the Rohde & Schwarz ZVA24 network analyzer.

Parameter	Performance
Frequency range	67 GHz
Port	4
Dynamic range	>150 dB
Output power	>18 dBm
Measurement speed	<2 μs
Sample time	430 ns
IF bandwidth	15 MHz

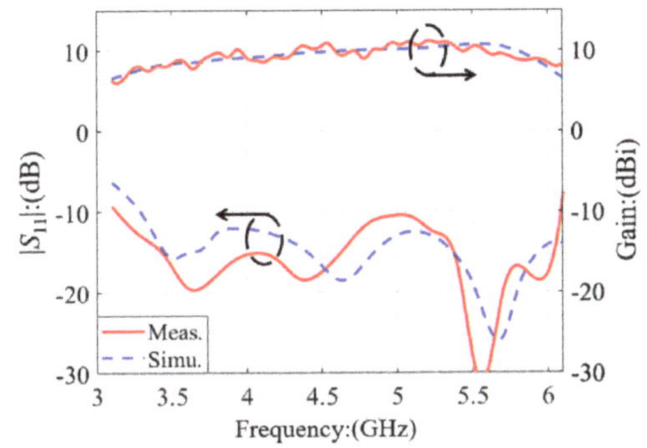

(**a**)

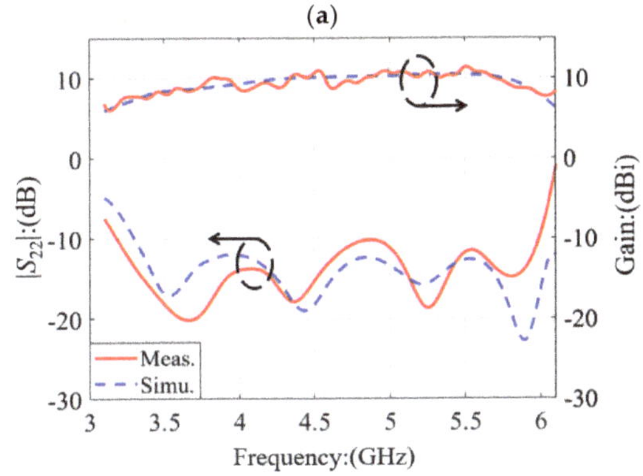

(**b**)

Figure 5. Measured and simulated *S*-parameters and gains: (**a**) Port 1; (**b**) Port 2.

The calculated and tested port isolations are plotted in Figure 6. Due to the assembly errors, there are certain deviations between the simulated and tested results. Though the tested $|S_{12}|$ is slightly different from the simulation result, the tested $|S_{12}|$ can be maintained below −31 dB using the optimized feeding network. So, satisfactory port isolation has been accomplished for the dual-polarized antenna.

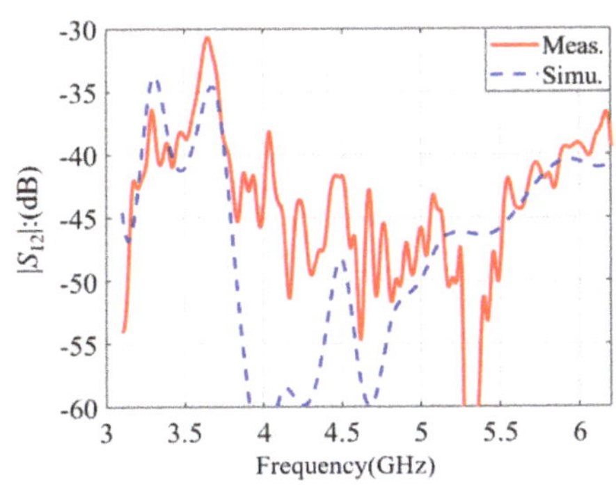

Figure 6. Calculated and tested port isolation.

The radiation performance was obtained in an anechoic chamber. An ATS200 multi-probe test system with a spherical near-field (SNF) testing method was used to measure the radiation properties of the antenna [33]. The near-field testing environment had 23 near-field probes in Figure 7. Its performance parameters are in Table 4. The antenna was placed at the center of the spherical measurement range, and the near-fields were tested at a series of points on a sphere enclosing the antenna. The SNF method is used to measure the far-field radiation properties of an antenna, which cannot be directly measured in a laboratory environment. The technique involves measuring the electromagnetic field on a spherical surface surrounding the antenna; then, an algorithm is used to convert the measurements into a far-field radiation pattern. The advantages of the SNF method include its ability to measure the far-field radiation properties of an antenna in a laboratory environment and its accuracy in measuring complex radiation patterns. The method is widely used in the design and testing of antennas for various applications.

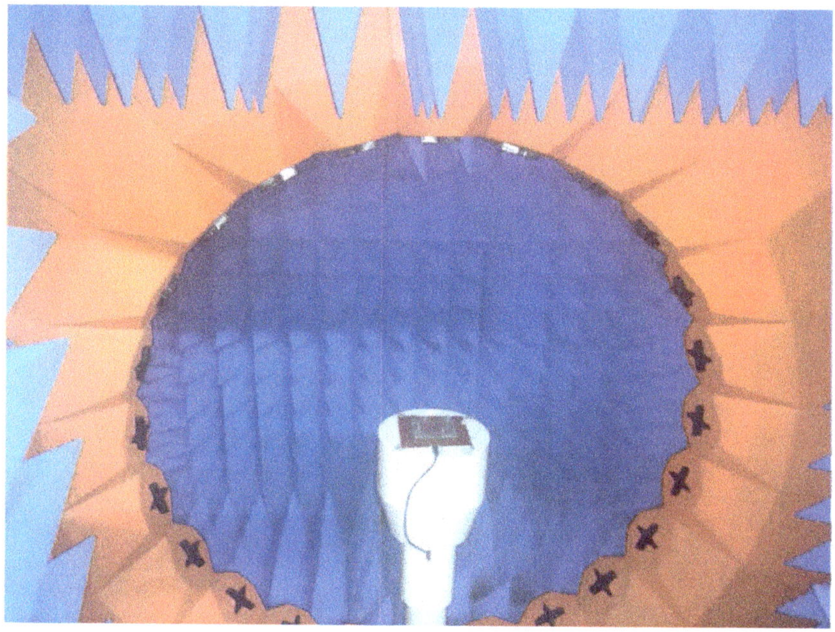

Figure 7. Antenna prototype testing scenario in the chamber.

Table 4. The performance parameters of the multi-probe test system.

Parameter	Performance
Frequency range	800 MHz~6 GHz
Dynamic range	75 dB
Gain stability	0.3 dB
TRP stability	0.5 dB
TIS stability	0.7 dB
Polarization	Circular, linear, elliptical

The tested gain performances are also plotted in Figure 5. For the 5G and 5G Wi-Fi bands between 3.3 GHz and 5.875 GHz, the measured gain of Port 1 varied between 7.9 dBi and 11.1 dBi. The average value was 9.62 dBi with a tested peak value of 11.3 dB at 5.2 GHz. The measured gain of Port 2 fluctuated between 7.7 dBi and 11.3 dBi. The average value was 9.64 dBi with a measured peak value of 11.3 dB at 5.5 GHz.

The calculated and tested radiation patterns of the two polarizations are plotted in Figures 8 and 9 at 4 GHz and 5 GHz. It is noticed that broadside radiation patterns have been achieved across the passband. The calculated and tested co-polarization curves almost overlap with each other. The tested cross-polarizations at 4 GHz and 5 GHz were less than −14.8 dB for Port 1 and −18 dB for Port 2. The measured back radiation levels at 4 GHz and

5 GHz were less than −11 dB for Port 1 and −10 dB for Port 2. So, low cross-polarization and back-lobe levels have been obtained.

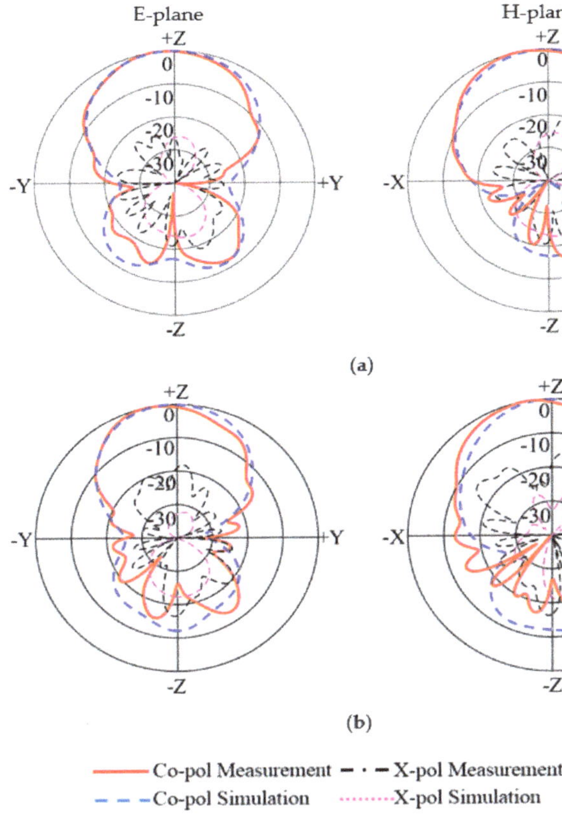

Figure 8. Calculated and tested radiation patterns of Port 1: (**a**) 4 GHz; (**b**) 5 GHz.

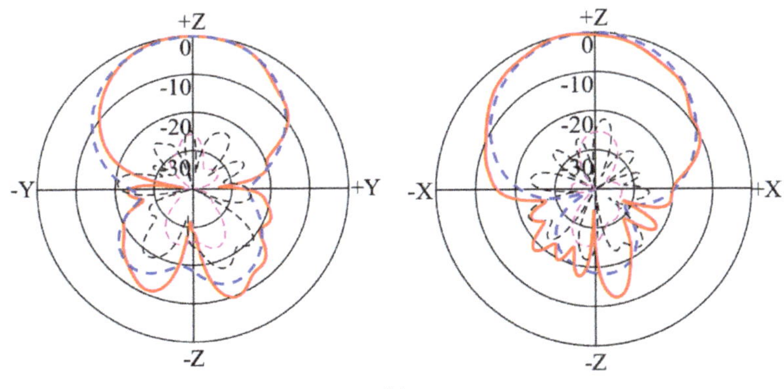

Figure 9. *Cont.*

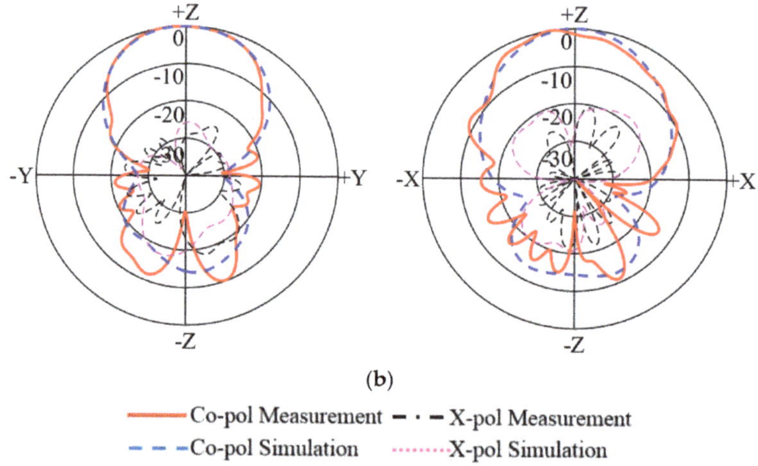

Figure 9. Calculated and tested radiation patterns of Port 2: (**a**) 4 GHz; (**b**) 5 GHz.

4. Discussions

The antenna's structures were discussed to clarify their influence on the antenna's performance, including the top layer and feedline. Many theories have been developed for antenna mode analysis; for example, the eigenmode [30] and characteristic mode [34]. In this design, the electric field distributions were analyzed to investigate the working mechanism. Port 2 was selected for the following discussion.

Firstly, the influence of the top layer's structure on the antenna performance will be discussed. The comparison of the *S*-parameters with and without the top layer is conducted in Figure 10. It is clear that the top layer had a significant effect on impedance matching. The curve of $|S_{22}|$ without the top layer was above −10 dB. So, the antenna with only the middle layer had poor impedance matching. After loading the top layer structure, the $|S_{22}|$ was improved effectively, and the curve was below −10 dB. So, a top layer structure can improve impedance matching effectively. It was concluded that the middle layer and the ground formed the basic microstrip structure and the top layer was loaded to augment the antenna's performance.

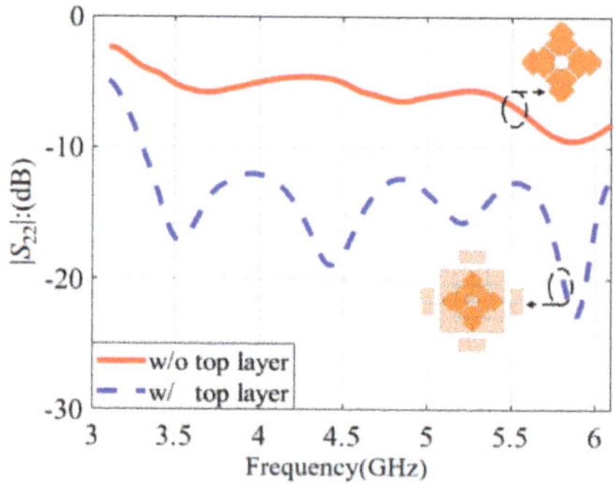

Figure 10. The influence of the top layer on the *S*-parameter.

The *S*-parameters with and without edge patches are compared in Figure 11. The red solid line is based on 16 patches while the blue dashed line is based on the proposed design with 8 edge patches. It is noticed that the impedance matching with 16 patches deteriorated slightly near 5.5 GHz and the $|S_{22}|$ was barely maintained at −10 dB. After loading the eight edge patches, the impedance at the higher frequency became better and the $|S_{22}|$ was less than −12 dB across the passband. So, edge patches can improve impedance matching at a higher band effectively, and a wider bandwidth can be accomplished.

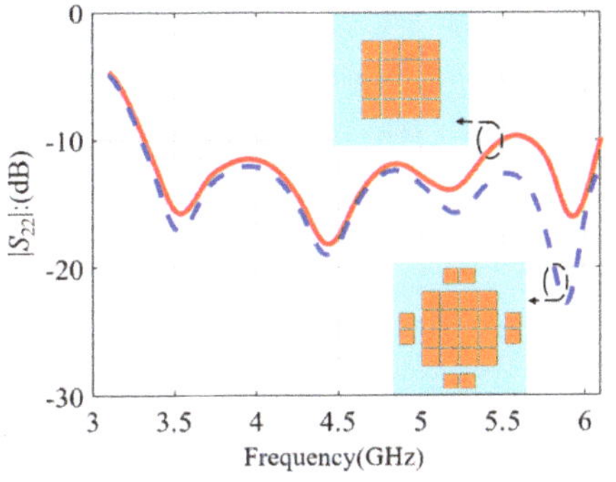

Figure 11. The influence of edge patches on the *S*-parameter.

The *S*-parameters with different feeding strips are compared in Figure 12. The red solid line is with a common straight feeding strip, while the blue dashed line is based on the feeding strip with a modified terminal structure. It is observed that the $|S_{22}|$ with an ordinary straight feeding strip was only −7 dB near 4 GHz. To optimize the working frequency band, the feeding strip structure was modified with an optimized terminal structure. Then, the impedance matching near 4 GHz was improved effectively and the $|S_{22}|$ became better than −12 dB across the passband. So, a modified feeding strip can improve the S-parameter at 4 GHz.

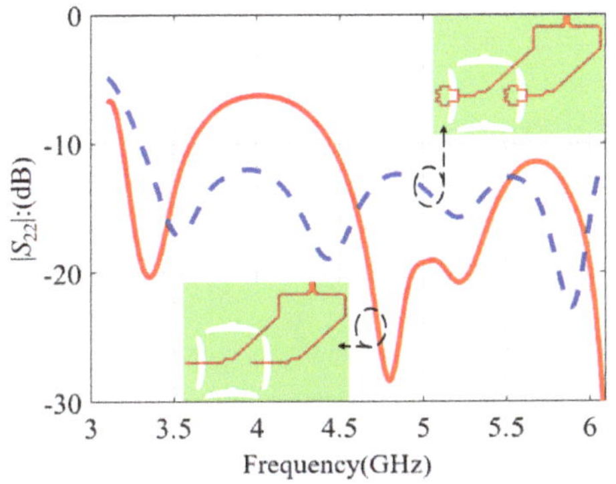

Figure 12. The influence of different feeding strips on the *S*-parameter.

To analyze the working mechanism of this dual-polarized antenna, effective and instantaneous electric field distributions between the radiating patches and the ground are plotted in Figures 13 and 14 with the electric field in the horizontal direction. Three frequency points were selected at 3.5 GHz, 4.5 GHz, and 5.5 GHz. The eigenmode method was adopted to discuss its working principle. It is observed from Figure 13 that the effective electric field distributions were symmetrical with four maximum points along the *Y*-axis. As the antenna had a large resonant structure at a low frequency, the effective electric field area was also relatively large in the polarization direction. When the frequency increased, the effective electric field area was reduced in the polarization direction.

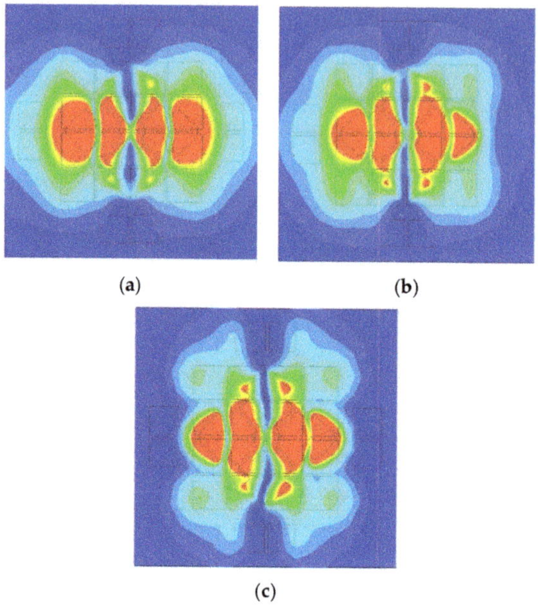

Figure 13. Effective electric field distributions of the microstrip antenna: (**a**) 3.5 GHz; (**b**) 4.5 GHz; (**c**) 5.5 GHz.

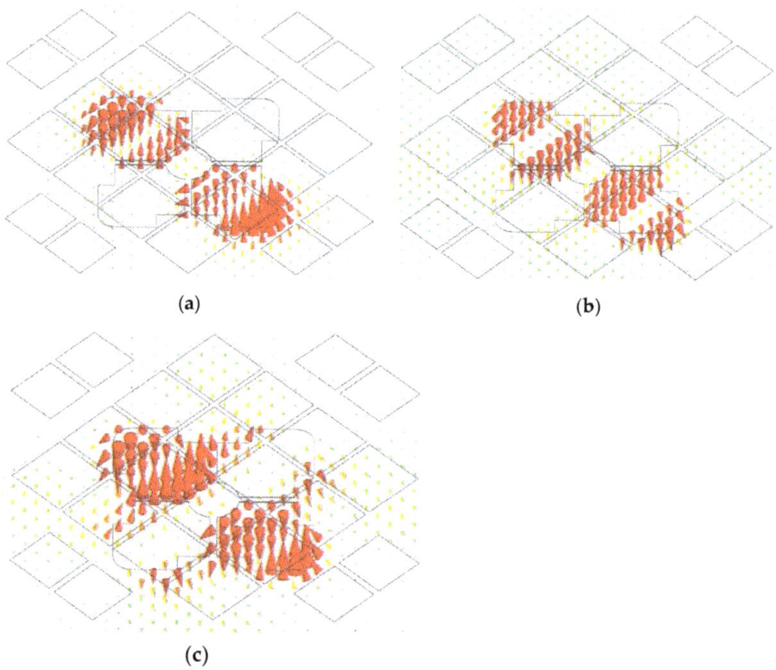

Figure 14. Instantaneous electric field distributions of the proposed antenna: (**a**) 3.5 GHz; (**b**) 4.5 GHz; (**c**) 5.5 GHz.

Corresponding to the effective electric field distribution, the instantaneous electric field is plotted in Figure 14. It is noticed that the electric field direction changed three times in the polarization direction at three frequencies. Because of these separated patch structures, the electric field phase changed by 180 degrees at the center in the polarization direction. Based on the electric field distribution, the proposed double-layer microstrip antenna was mainly in anti-phase TM_{20} mode at low, middle, and high frequencies. So, wideband performance was obtained based on the proposed structure.

5. Conclusions

A dual-polarized double-layer antenna was investigated for 5G and 5G Wi-Fi. The middle layer and the ground constructed the basic antenna structure, and a metasurface structure was applied for the top layer to improve impedance matching. Broadband performances of 64.1% (3.13 GHz~6.08 GHz) and 61.1% (3.18 GHz~5.98 GHz) were achieved with anti-phase TM_{20} modes across the passband. A feeding strip with a modified terminal structure can improve the impedance matching of the middle band, and parasitic edge patches can optimize the impedance matching of the higher band effectively. A high port isolation of 31 dB was achieved. Furthermore, the antenna had a low profile of $0.096\lambda_0$ for space saving and easy installation. The antenna element had realized high peak gains of 11.1 dB and 11.3 dB with broadside radiation patterns and low cross-polarizations. With these favorable characteristics, this broadband dual-polarized antenna should find widespread applications for 5G and 5G Wi-Fi communications.

Author Contributions: Conceptualization, W.A.; methodology, W.A.; software, X.T.; validation, W.A., X.T. and S.W.; formal analysis, J.W.; investigation, W.A., X.T. and S.W.; resources, W.A. and J.W.; data curation, W.A., X.T. and S.W.; writing—original draft preparation, W.A. and X.T.; writing—review and editing, W.A. and J.W.; visualization, W.A., X.T. and S.W.; supervision, J.W.; project administration, W.A. and J.W.; funding acquisition, J.W. All authors have read and agreed to the published version of the manuscript.

Funding: This research was funded by the State Key Laboratory of Complex Electromagnetic Environment Effects on Electronics and Information System (no. CEMEE2022G0201), in part by the National Natural Science Foundation of China (grant no. 61701339), in part by the National Key Research and Development Program of China (Nano Science and Technology Project, no. 2016YFA0202200), and in part by the AoShan Talents Cultivation Program Supported by Qingdao National Laboratory for Marine Science and Technology under grant no. 2017ASTCP-OS03.

Data Availability Statement: Not applicable.

Conflicts of Interest: The authors declare no conflict of interest.

References

1. Sun, W.; Li, Y.; Zhang, Z.; Chen, P.-Y. Low-profile and wideband microstrip antenna using quasi-periodic aperture and slot-to-CPW transition. *IEEE Trans. Antennas Propag.* **2019**, *67*, 632–637. [CrossRef]
2. Sun, W.; Li, Y.; Zhang, Z.; Feng, Z. Broadband and low-profile microstrip antenna using strip-slot hybrid structure. *IEEE Antennas Wirel. Propag. Lett.* **2017**, *16*, 3118–3121. [CrossRef]
3. An, W.; Li, Y.; Fu, H.; Ma, J.; Chen, W.; Feng, B. Low-profile and wideband microstrip antenna with stable gain for 5G wireless applications. *IEEE Antennas Wirel. Propag. Lett.* **2018**, *17*, 621–624. [CrossRef]
4. Yeap, S.B.; Chen, Z.N. Microstrip patch antennas with enhanced gain by partial substrate removal. *IEEE Trans. Antennas Propag.* **2010**, *58*, 2811–2816. [CrossRef]
5. An, W.; Shen, Z.; Wang, J. Compact low-profile dual-band tag antenna for indoor positioning systems. *IEEE Antennas Wirel. Propag. Lett.* **2017**, *16*, 400–403. [CrossRef]
6. Liu, N.-W.; Zhu, L.; Choi, W.-W.; Zhang, X. Wideband shorted patch antenna under radiation of dual-resonant modes. *IEEE Trans. Antennas Propag.* **2017**, *65*, 2789–2796. [CrossRef]
7. Lu, W.; Li, Q.; Wang, S.; Zhu, L. Design approach to a novel dual-mode wideband circular sector patch antenna. *IEEE Trans. Antennas Propag.* **2017**, *65*, 4980–4990. [CrossRef]
8. An, W.X.; Li, S.R.; Sun, W.Y.; Li, Y. Low-profile wideband microstrip antenna based on multiple modes with partial apertures. *IEEE Antennas Wirel. Propag. Lett.* **2019**, *18*, 1706–1713. [CrossRef]

9. Mirmozafari, M.; Zhang, G.; Saeedi, S. A dual linear polarization highly isolated crossed dipole antenna for MPAR application. *J. IEEE Antennas Wirel. Propag. Lett.* **2017**, *16*, 1879–1882. [CrossRef]
10. Wu, Q.; Liang, P.; Chen, X. A broadband ±45° dual-polarized multiple-input multiple-output antenna for 5G base stations with extra decoupling elements. *J. Commun. Inf. Netw.* **2018**, *3*, 31–37. [CrossRef]
11. Ta, S.X.; Nguyen, M.D.; Nguyen, T.H.Y. A low-profile ± 45° dual-polarized antenna based on metasurface and its arrays for base station applications. *J. J. Electromagn. Waves Appl.* **2018**, *32*, 1748–1767. [CrossRef]
12. Li, Y.; Wang, C.; Yuan, H.; Liu, N.; Zhao, H.; Li, X. 5G MIMO antenna manufactured by 3-D printing method. *IEEE Antennas Wirel. Propag. Lett.* **2017**, *16*, 657–660. [CrossRef]
13. Siu, L.; Wong, H.; Luk, K.M. A dual-polarized magneto-electric dipole with dielectric loading. *IEEE Trans. Antennas Propag.* **2009**, *57*, 616–623. [CrossRef]
14. Wu, B.Q.; Luk, K.M. A broadband dual-polarized magneto-electric dipole antenna with simple feeds. *IEEE Antennas Wirel. Propag. Lett.* **2009**, *8*, 60–63.
15. Huang, H.; Liu, L. A low-profile, single-ended and dual-polarized patch antenna for 5G application. *IEEE Trans. Antennas Propag.* **2020**, *68*, 4048–4053. [CrossRef]
16. Sim, C.Y.D.; Chang, C.C.; Row, J.S. Dual-feed dual-polarized patch antenna with low cross polarization and high isolation. *IEEE Trans. Antennas Propag.* **2009**, *57*, 3321–3324. [CrossRef]
17. Wen, L.H.; Gao, S.; Luo, Q. A low-cost differentially driven dual-polarized patch antenna by using open-loop resonators. *J. IEEE Trans. Antennas Propag.* **2019**, *67*, 2745–2750. [CrossRef]
18. Deng, C.; Li, Y.; Zhang, Z. A wideband high-isolated dual-polarized patch antenna using two different balun feedings. *J. IEEE Antennas Wirel. Propag. Lett.* **2014**, *13*, 1617–1619. [CrossRef]
19. Yang, X.; Ge, L.; Wang, J. A differentially driven dual-polarized high-gain stacked patch antenna. *J. IEEE Antennas Wirel. Propag. Lett.* **2018**, *17*, 1181–1185. [CrossRef]
20. Tang, Z.; Liu, J.; Cai, Y.M. A wideband differentially fed dual-polarized stacked patch antenna with tuned slot excitations. *J. IEEE Trans. Antennas Propag.* **2018**, *66*, 2055–2060. [CrossRef]
21. Liu, Y.; Wang, S.H.; Wang, X.D.; Jia, Y.T. A differentially fed dual-polarized slot antenna with high isolation and low profile for base station application. *IEEE Antennas Wirel. Propag. Lett.* **2019**, *18*, 303–307. [CrossRef]
22. Lian, R.; Wang, Z.; Yin, Y. Design of a low-profile dual-polarized stepped slot antenna array for base station. *J. IEEE Antennas Wirel. Propag. Lett.* **2016**, *15*, 362–365. [CrossRef]
23. Wong, S.W.; Sun, G.H.; Zhu, L. Broadband dual-polarization and stable-beamwidth slot antenna fed by U-shape microstrip line. *J. IEEE Trans. Antennas Propag.* **2016**, *64*, 4477–4481. [CrossRef]
24. Alibakhshikenari, M.; Virdee, B.S.; See, C.H.; Shukla, P.; Moghaddam, S.M.; Zaman, A.U.; Shafqaat, S.; Akinsolu, M.O.; Liu, B.; Yang, J.; et al. Dual-polarized highly folded bowtie antenna with slotted self-grounded structure for sub-6 GHz 5G applications. *IEEE Trans. Antennas Propag.* **2022**, *70*, 3028–3033. [CrossRef]
25. Xue, K.; Yang, D.; Guo, C.; Zhai, H.; Li, H.; Zeng, Y. A dual-polarized filtering base-station antenna with compact size for 5G applications. *IEEE Antennas Wirel. Propag. Lett.* **2020**, *19*, 1316–1320. [CrossRef]
26. Li, Y.; Zhao, Z.; Tang, Z.; Yin, Y. Differentially fed, dual-band dual-polarized filtering antenna with high selectivity for 5G sub-6 GHz base station applications. *IEEE Trans. Antennas Propag.* **2020**, *68*, 3231–3236. [CrossRef]
27. Hou, Y.; Shao, Z.; Zhang, Y.; Mao, J. A wideband differentially fed dual-polarized laminated resonator antenna. *IEEE Trans. Antennas Propag.* **2021**, *69*, 4148–4153. [CrossRef]
28. Liu, W.; Chen, Z.N.; Qing, X. Metamaterial-based low-profile broadband mushroom antenna. *IEEE Trans. Antennas Propag.* **2014**, *62*, 1165–1172. [CrossRef]
29. Nie, N.-S.; Yang, X.-S.; Chen, Z.N.; Wang, B.-Z. A low-profile wideband hybrid metasurface antenna array for 5G and WiFi systems. *IEEE Trans. Antennas Propag.* **2020**, *68*, 665–671. [CrossRef]
30. Lin, F.H.; Chen, Z.N. Resonant metasurface antennas with resonant apertures: Characteristic mode analysis and dual-polarized broadband low-profile design. *IEEE Trans. Antennas Propag.* **2021**, *69*, 3512–3516. [CrossRef]
31. Sun, W.; Li, Y. Dual-polarized grid-slotted microstrip antenna with enhanced bandwidth and low profile. *ACES J.* **2019**, *34*, 451–456.
32. An, W.X.; Wang, S.S.; Luo, Y. Low-profile wideband dual-polarized microstrip antenna element for 5G communication. *Microw. Opt. Technol. Lett.* **2022**, *64*, 351–357. [CrossRef]
33. ATS2000 Multi Probe Testing System. Available online: https://www.xh-ioe.com/h-pd-12.html (accessed on 1 March 2023).
34. Suo, M.; Xiong, H.; Li, X.K. A flexible transparent absorber bandwidth expansion design based on characteristic modes. *J. Results Phys.* **2023**, *46*, 106265. [CrossRef]

Disclaimer/Publisher's Note: The statements, opinions and data contained in all publications are solely those of the individual author(s) and contributor(s) and not of MDPI and/or the editor(s). MDPI and/or the editor(s) disclaim responsibility for any injury to people or property resulting from any ideas, methods, instructions or products referred to in the content.

micromachines

Article

Design and Fabrication of a Low-Cost, Multiband and High Gain Square Tooth-Enabled Metamaterial Superstrate Microstrip Patch Antenna

Khaled Aliqab [1,*], Sunil Lavadiya [2], Meshari Alsharari [1], Ammar Armghan [1,*], Malek G. Daher [3] and Shobhit K. Patel [4]

1 Department of Electrical Engineering, College of Engineering, Jouf University, Sakaka 72388, Saudi Arabia
2 Department of Information and Communication Technology, Marwadi University, Rajkot 360003, India
3 Physics Department, Islamic University of Gaza, Gaza P.O. Box 108, Palestine
4 Department of Computer Engineering, Marwadi University, Rajkot 360003, India
* Correspondence: kmaliqab@ju.edu.sa (K.A.); aarmghan@ju.edu.sa (A.A.); Tel.: +96-6557202146

Abstract: The manuscript represents a novel square tooth-enabled superstrate metamaterial loaded microstrip patch antenna for the multiple frequency band operation. The proposed tooth-based metamaterial antenna provides better gain and directivity. Four antenna structures are numerically investigated for the different geometry of the patch and tooth. These proposed structures are simulated, fabricated, measured, and compared for the frequency range of 3 GHz to 9 GHz. The electrical equivalent model of the split-ring resonator is also analyzed in the manuscript. The comparative analysis of all of the proposed structures has been carried out, in terms of several bands, reflectance response, VSWR, gain and bandwidth. The results are compared with previously published works. The effects are simulated using a high-frequency structure simulator tool with the finite element method. The measured and fabricated results are compared for verification purposes. The proposed structure provides seven bands of operation and 8.57 dB of gain. It is observed that the proposed design offers the multiple frequency band operation with a good gain. The proposed tooth-based metamaterial antenna suits applications, such as the surveillance radar, satellite communication, weather monitoring and many other wireless devices.

Keywords: microstrip patch antenna; metamaterial; advanced; multiband; directivity; gain

Citation: Aliqab, K.; Lavadiya, S.; Alsharari, M.; Armghan, A.; Daher, M.G.; Patel, S.K. Design and Fabrication of a Low-Cost, Multiband and High Gain Square Tooth-Enabled Metamaterial Superstrate Microstrip Patch Antenna. *Micromachines* **2023**, *14*, 163. https://doi.org/10.3390/mi14010163

Academic Editors: Trushit Upadhyaya and Hari Shankar Singh

Received: 14 December 2022
Revised: 3 January 2023
Accepted: 5 January 2023
Published: 8 January 2023

Copyright: © 2023 by the authors. Licensee MDPI, Basel, Switzerland. This article is an open access article distributed under the terms and conditions of the Creative Commons Attribution (CC BY) license (https://creativecommons.org/licenses/by/4.0/).

1. Introduction

An antenna is a prime component in all kinds of wireless communication. We are converting most wired technology to wireless, to ease equipment usage and to cover long-range communications, due to the heavy demand for multiple applications, such as mobile communication, wireless local area networks, global positioning system (GPSs), radio frequency identification (RFID) and many more. RF engineers always face challenges, such as multi-bands, wide bandwidths, high gains, and power-efficient antennas [1]. There is a need to target multiple applications with the same device. To achieve the same multiband and broadband antenna, it is helpful to cover the broader spectrum. An antenna with a low profile, a small size and broadband should be located in the front of the system to target multiple wireless communication applications [2]. One of the significant challenging parts is to design a smaller antenna, as a microstrip patch antenna is more suitable. A microstrip patch antenna is also called a printed antenna or patch antenna [3]. The advantages of patch antennas are their low cost, better reconfigurability and ease of fabrication [4]. However, due to certain drawbacks of microstrip patch antennas, such as their low gain, low bandwidth and low directivity, some improvement is required [5,6]. Multiband antennas have tremendous applications in mobile communication [7,8].

There is always controversy in bandwidths and antenna sizes. The rise of one parameter will degrade another parameter. There are many ways available to achieve a multiband operation, such as engraving slots on a patch [9], loading shorted slots with outers [10], etching unit-cells of metamaterial on/around the patch and loading it on a substrate, ground layer [11]. The explosion in cutting-edge wireless communication methods has led to a steady rise in the demand for compact mobile devices. Therefore, new approaches to the design of wireless components are needed to meet many performance requirements at once. Every wireless mobile component's design and performance improvement currently faces significant hurdles, including the need for a compact size, lightweight, low profile and cheap cost. The antenna is one of the wireless components that will need to be improved, to keep up with the demands of modern communication networks.

Many methods are available for enhancing the bandwidth and the gain of an antenna. It can be accomplished by raising the thickness of the substrate [12], but the surface radiation is affected by changing substrate's thickness [13]. The solution is achieved by using a substrate with a low profile [14], a different impedance matching network and a slot antenna geometry [8,15]. A novel meandered high-impedance patch antenna provides the desired multiband behavior by inserting narrow slits in the patch [16]. A broadband folded patch antenna is used for wireless local area network applications [17]. Multiple patches and ground plane slots are used to achieve the multiband operation [18]. Different shapes in the patch also enrich the antenna's bandwidth [19,20]. There are some approaches available for the antenna's gain improvement, such as the partial ground method, the diffractive ground [21], an artificially soft surface [22], micromachining technology and a substrate with a high dielectric constant, such as liquid (water-sea), rather than copper. However, there is a high fabrication cost for anElectronic Band Gap (EBG), including the artificially soft surfaces and micromachining technology. The substrate multilayer provides a better bandwidth but degrades the antenna's gain, as well as the efficiency of antenna's structure [23]. The high gain antenna using the hexagon-shaped metamaterial elements, is presented in the article [6]. The shape variation of the metamaterial elements affects the band response [24,25]. The metamaterial concept-inspired wearable fractal antenna helps to target different IoT applications [26]. The fractal Sierpinski-shaped antenna concept helps to target multiple wireless applications with miniaturization features [27].

A superstrate provides an effective way to reduce the mutual coupling between radiating elements. It reduces the turns' mutual coupling and supports the antenna's directivity improvement [28,29]. There are many methods available for the antenna's improvement, such as an array of the antenna, a surface-mounted horn antenna, composite conductors and a lens antenna [30–32]. At the same time, the main limitations of these designs are the large space requirements and the hardware needed for the high-gain antennas [6,33]. The listed problems are solved by adding artificial properties to the material, these are called metamaterials.

Metamaterial provides a negative value of permittivity (ε) and permeability (μ) [34]. The primary elements for making metamaterial are the complementary split-ring resonators (CSRRs) and metal wires. The gap between the two terminals of a metal wire behaves as the capacitance, and their curvature shape provides an inductance effect. The size of such a structure is less than the ordinary resonating structure [35]. The metamaterial approach helps to achieve targets, due to their negative permittivity and negative permeability concept [36]. Metamaterial-based designs are used for making perfect lenses, invisibility cloaks, electromagnetic bandgaps (EBGs) and photonic bandgaps (PBGs) [37]. In antenna development, the composite electromagnetic bandgap structure (EBG) and the metamaterial inhibit the electromagnetic wave for a specific range of frequency [38,39]. This structure innovates a new way of developing compact and better performing RF components [40]. The two-dimensional EBG design confines the surface wave in a patch structure and metamaterial, enhancing the antenna's gain, directivity and reflectance response [41]. There are a few possible ways for the miniaturization of metamaterial antennas. First is a substrate with a high permittivity, a low tangent loss and a rising thickness. Second is a zero-impedance plane, similar to the perfect electric conductor. Third is inserting an infinite impedance identical to an excellent magnetic conductor. Efficiency, in the third case, is reduced due to the dissipative loss [42,43]. Fourth is using metamaterial above a patch antenna to condense the surface waves. An array of metamaterial rings will enrich the antenna's gain, but it also increases the size of the antenna. The solution is provided by inserting a tooth on the exterior surface of the ring [44].

The manuscript presented here represents a novel metamaterial-based multi-layered superstrate structure, to achieve the multiband operation for S, C and X frequency bands. Gain improvement has been achieved by adding a tooth to the metamaterial structure [45]. Four designs are simulated to achieve a high gain, multiband and broadband characteristics. The first design is a patch with connectors and metamaterial rings with the tooth. The second is a patch with connectors and metamaterial without a tooth. The third is a patch without connectors and metamaterial with the tooth. Fourth is a patch without connectors and metamaterial without a tooth. The charge distribution in the patch region is affected, due to the connector section of the exterior and interior patch elements. Therefore, it leads to a change in the reflectance response plot. The effect of the tooth over the metamaterial loaded structure helps to attain a better performance. All designs are compared using different substrate materials, such as Rogers RT Duroid 5880 and FR4 materials. The simulated results are compared, in terms of the number of bands, the bandwidth, VSWR, the reflectance response (S_{11}) and gain.

2. Materials and Methods

A 3D view of the proposed superstrate rectangular microstrip patch antenna is shown in Figure 1. The patch antenna has one rectangular cut/cropped section, which helps to achieve a multiband response and a lower reflectance response (S_{11}). Two substrate layers of dielectric materials are used to achieve a broader bandwidth. A novel thing introduced in the manuscript is a tooth around the split ring resonator. The design parameters of the rectangular patch antenna, include the length (L) and width (w) of the ground plane and the patch are calculated, based upon selecting the height of the substrate (h), the dielectric constant of the substrate (ε_r) and the resonating frequency (f_r) [46].

Figure 2 represents the fabricated prototype of the proposed antenna structure. Figure 2a represents a split ring resonator with the tooth. Figure 2b shows a split ring resonator without a tooth. Figure 2c displays the disconnected inner and exterior sides of the patch. Figure 2d represents the connected inner and exterior parts of the patch. Figure 2e,f show the side views of the proposed structure. Figure 2g shows the anechoic chamber during the antenna directivity measurement. Figure 2f displays the reflectance response measurement using a vector network analyzer.

The side view of the designs are represented in Figure 3a. The ground layer, the patch, the split ring resonator, and the tooth thickness are 0.35 mm. The height of the substrate layer is 1.5 mm. The dimensions of the rectangular ground layer and the substrates are 66.4 mm. The patch and all of the split-ring resonators are designed in a centered position. The upper view of the antenna is represented in Figure 3b. The first (exterior), second (middle), and third (interior) split-ring resonators are positioned at 8.2 mm, 15.3 mm, and 22.2 mm, from the exterior border, respectively. The width of the SRR is 2 mm. The size of the tooth is 1.5 × 1.5 mm^2. The gap between the two terminals of the SRR is 2 mm, and it is placed 32.2 mm away from the exterior border. The distance between the tooth located in the split-ring resonators is, respectively, 4.44 for the first (exterior), 3.78 mm for the second (middle) and 3.4 mm for the third (interior). The top view of the patch is shown in Figure 3c. The patch is kept 5 mm away from the exterior edge. The rectangular patch outer structure (POS) dimension is 56.4 mm. The cropped/cut section in the patch is 2mm. The patch interior structure (PIS) is 46 mm. The connectors for shorting the PIS with the POS are 2 mm. The connecters are kept 32.2 mm from the antenna's exterior. The antenna's structure is excited by applying the input from a coaxial feed. The coaxial feed is kept 52.8 mm by 13.5 mm away from the exterior. The radius of the coaxial interior is 0.12 mm, and the exterior is 0.43 mm. The ground layer, the patch, the SRR, the tooth and coaxial's interior are made of copper. The coaxial's exterior is made of plastic. Two types of substrate materials are considered Rogers RT Duroid 5880 (Ɛ—2.2) and FR4 (Ɛ—4.4) [47].

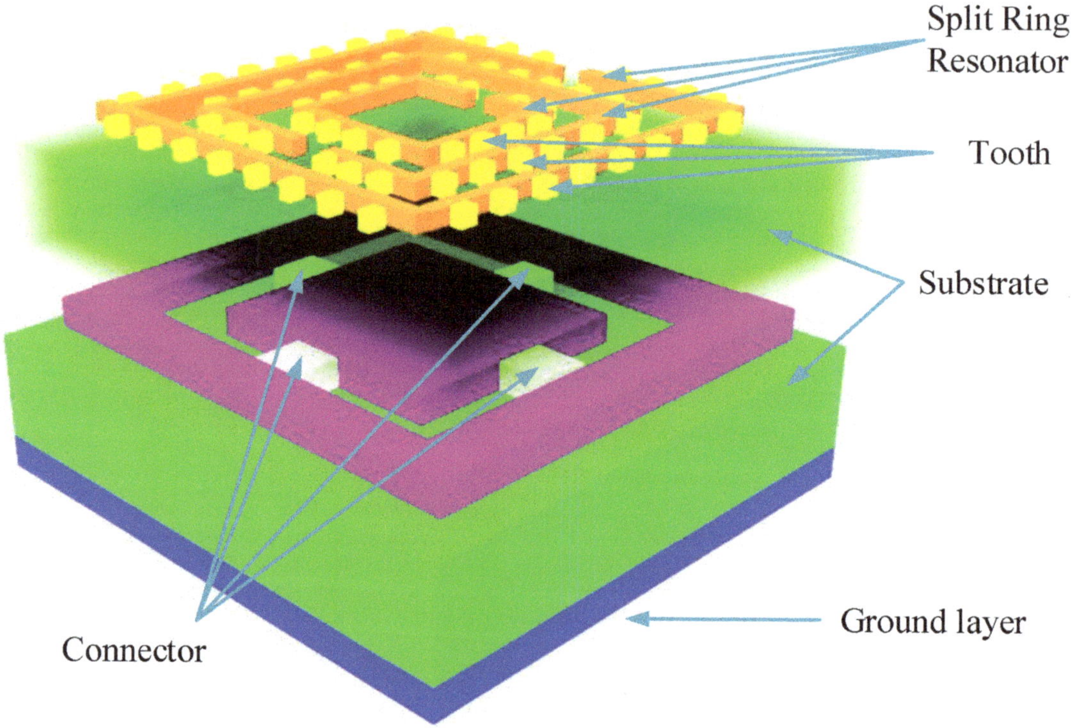

Figure 1. Three-dimensional interpretation of the proposed rectangular microstrip patch antenna. Two superstrate layers of substrates are used. The patch consists of a rectangular cut/cropped slot. Three SRRs with the tooth are located at the top of the upper substrate.

In a SRR, the inductance effect is induced by the circular shape and the capacitance by the space is between two ring terminals. The inductance (Ls) and capacitance (C_s) per unit length can be calculated by Equations (1) and (2) [48].

$$Ls = \frac{\mu_0 b}{\sqrt{\pi}}\left[\mathrm{Log}\left(\frac{32b}{w\sqrt{\pi}}\right) - 2\right] \tag{1}$$

$$C_s = \varepsilon\frac{wt}{2g} \tag{2}$$

where, μ_0 is the free space permittivity ($\mu_0 = 4\,\pi \times 10^{-7}$ N/A^2), the width of the ring is w, g is a gap between two split rings, b is the ring length, the series capacitance is C_s, t is the ring thickness. The RLC circuit of the SRR is shown in Figure 4. The resonating frequency of the proposed design can be calculated using Equation (3) [49].

$$f = \frac{1}{2\pi\sqrt{L_s C_s}} \tag{3}$$

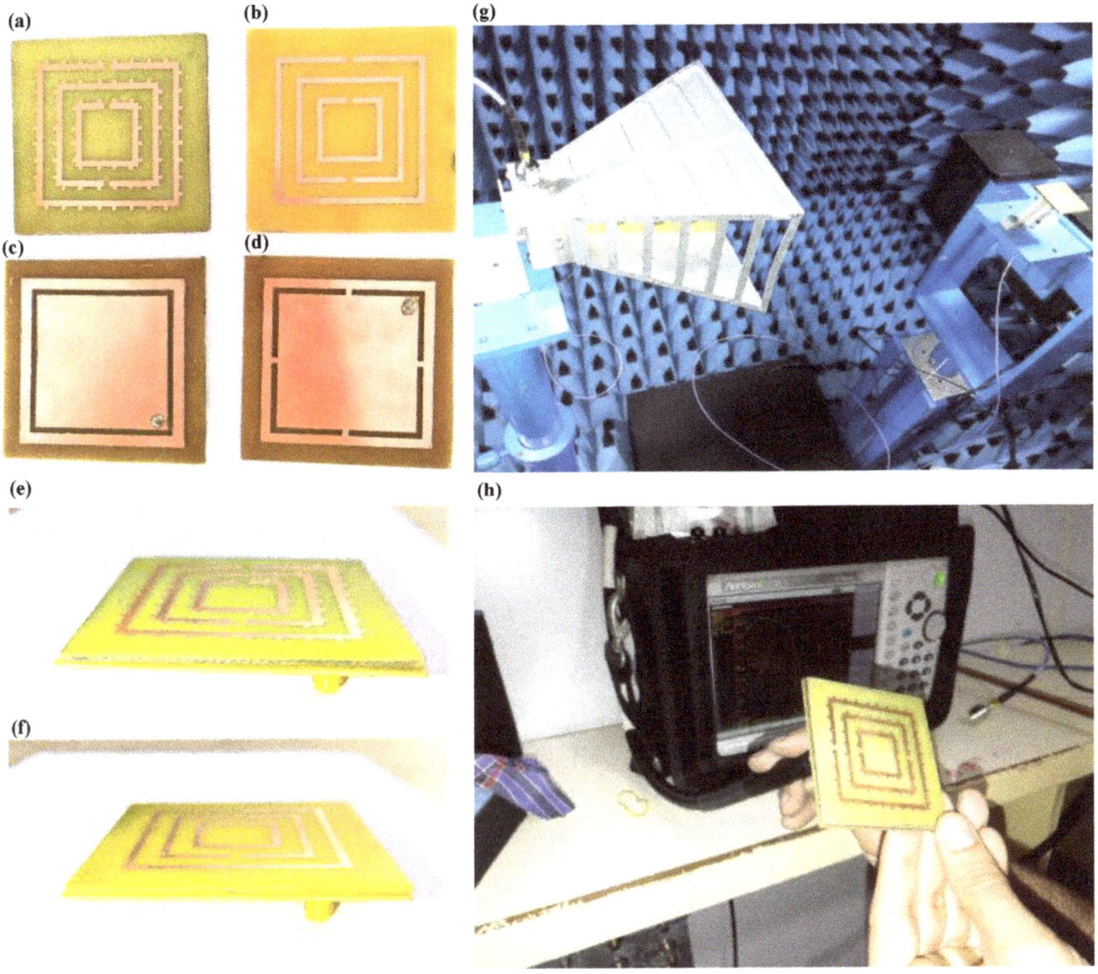

Figure 2. (**a–f**) The fabricated prototype of the proposed antenna structure. (**e**) Directivity measurement using the anechoic chamber. (**f**) Reflectance response measurement using a vector network analyzer. (**g**) Testing of antenna in Anachoic chamber. (**h**) Measuring reflectance response using Vector Network Analyser.

The performance index of the high frequency operated antenna structure can be calculate using the S parameters. The impedance and refractive index is calculated using the reflectance (S_{11}) and the transmittance (S_{21}) response using in Equations (4) and (5).

$$n = \frac{1}{kd}\cos^{-1}\left[\frac{1}{2S_{21}}X\left(1 - S_{11}^2 + S_{21}^2\right)\right] \tag{4}$$

$$z = \sqrt{\frac{(1 + S_{11}^2) - (S_{21}^2)}{(1 - S_{11}^2) - (S_{21}^2)}} \tag{5}$$

where, d is the substrate width, n is the refractive index, k is the wave vector and z is the wave impedance.

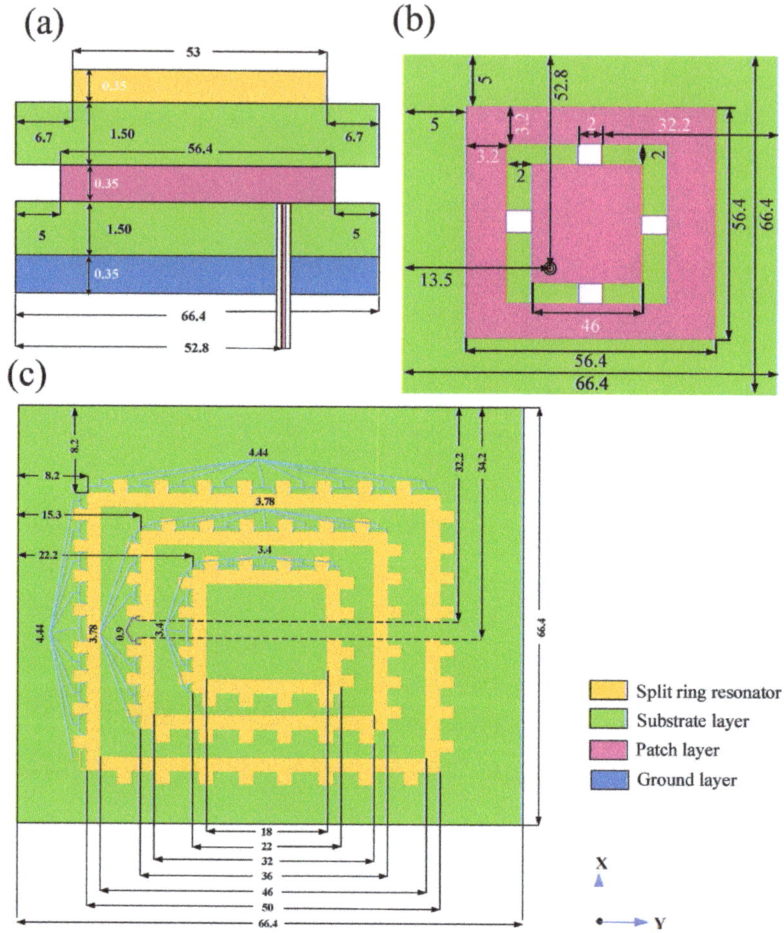

Figure 3. (**a**) Side view of a presented structure. (**b**) Top view of the patch. (**c**) Top view of the rectangular shaped MPA loaded with the tooth and SRR. Dimensions represented in the figure are in mm.

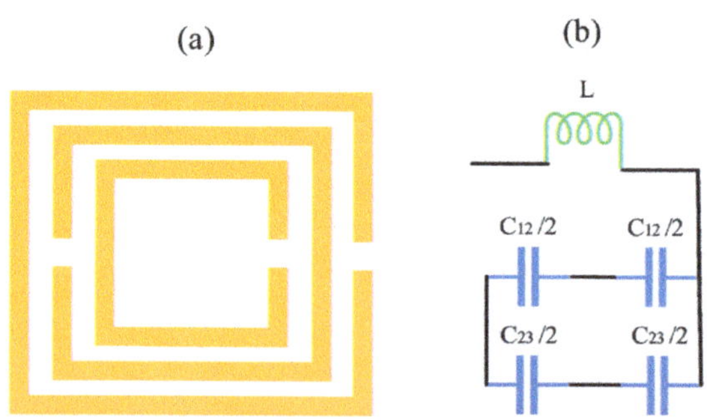

Figure 4. (**a**) The complementary SRR is represented. (**b**) Equivalent model of the structure.

3. Result and Discussion

The proposed antenna design works in the range of the *S*, *C*, and *X* frequency bands. Therefore, the response analysis should be carried out by the scattering parameters (S-Parameters). The S-parameters represent the electrical behavior of a linear electrical system, describing the input and output relations among the ports of an electrical system. Unambiguously, for the high frequency, it is essential to represent a given network, in terms of waves, not by the voltage (v) or current (i).

Figure 5 represents four antenna structures that are simulated, based upon the on and off in the interior-exterior of the patch and the tooth in the split ring resonators. The following notations were given for the design of the antenna: the connected patch interior and patch outer(CPIO), the disconnected patch interior and patch exterior (DCPIO), the SRR with a tooth (SWT), and the split ring resonator without a tooth (SWOT). The four combinations are represented in Table 1. The first structure is the CPIO and SWT. The second structure is the CPIO and SWOT. The third structure is the DCPIO and SWT. The fourth structure is the DCPIO and SWOT. The design of all of these structures is represented in Figure 6. The performance of the four designs is observed by changing both the substrate materials of Rogers RT Duroid 5880 and FR4.

S_{11} or gamma represents the amount of waves radiated by the antenna. The response is observed for the different frequency values. S_{11} is chosen for values less than -10 dB. The substrate material varies for all four structures, and the performance is analyzed and compared. The reflectance response is represented by S_{11}. The resonating behavior of the antenna's structure at a particular frequency is described by S_{11}. How closely an antenna's impedance matches that of the radio or transmission line into the load may be quantified using the voltage standing wave ratio (VSWR) value. The reflectance response and the VSWR are both directly proposed for each other. The cross-verification among both responses gives a better clarity of the proposed design. The wideband antenna transfers information over a broad range of the frequency spectrum, whereas the narrowband signals occupy a considerably smaller fraction of the spectrum and need less transmitting power for a given application. Tactical military radios, industrial monitoring, shorter-range fixed-location wireless applications, radio-frequency identification, and commercial vehicle remote keyless entry devices are all types of uses that have traditionally relied on narrowband antennas to achieve reliable links in varying operating environments. Likewise, cellular communication networks use several very narrow bands to provide a wide range of service applications. The effect of noise is also limited, due to the usage of a narrow band, compared to the wideband response.

Figure 6 represents an analysis of S_{11} for the simulated and measured analysis, using FR-4 material as the substrate. The similarity is observed in the measured and simulated results. There are three frequency bands observed in the CPIO and SWT antenna structures. The value of the reflectance response achieved is -26.11 dB at a resonance frequency of 3.38 GHz, and the maximum bandwidth achieved for this is 0.22 GHz. There are two frequency bands observed in the CPIO and SWOT antenna structures. The reflectance response value reached is -21.52 dB at a resonance frequency of 3.39 GHz, and the maximum bandwidth achieved for this is 0.07 GHz. There are three frequency bands observed in the DCPIO and SWT antenna structures. The value of the reflectance response achieved is -11 dB at a resonance frequency of 7.27 GHz, and the maximum bandwidth achieved for this is 0.24 GHz. There is one frequency band observed in the DCPIO and SWOT antenna structures. The reflectance response value reached is -10.10 dB at a resonance frequency of 8.64 GHz, and the maximum bandwidth achieved for this is 0.02 GHz.

Figure 7 represents a S_{11} plot by choosing the substrate as Rogers RT Duroid 5880. There are seven frequency bands observed in the CPIO and SWT antenna structures. The maximum value of the reflectance response achieved is -33.79 dB at a resonance frequency of 6.49 GHz, and the maximum bandwidth achieved for this is 0.18 GHz. There are six frequency bands observed in the CPIO and SWOT antenna structures. The maximum value of the reflectance response achieved is -34.54 dB at a resonance frequency of 6.52 GHz, and the maximum bandwidth achieved for this is 0.16 GHz. There are three frequency bands observed in the DCPIO and SWT antenna structures. The maximum value of the reflectance response achieved is −13.79 dB at a resonance frequency of 8.83 GHz, and the maximum bandwidth achieved for this is 0.51 GHz. One frequency band is observed in the DCPIO and SWOT antenna structures. The reflectance response value reached is −10.05 dB at a resonance frequency of 8.80 GHz, and the maximum bandwidth achieved for this is 0.07 GHz. Data are represented in Table 2. It is observed that more bands and minimum reflectance responses are better in CPIO and SWT antenna structures in both types of the substrate-based design. The Rogers RT Duroid-based antenna structure provides more bands and a better reflectance response than the FR4-based substrate design. Data are represented in Table 3.

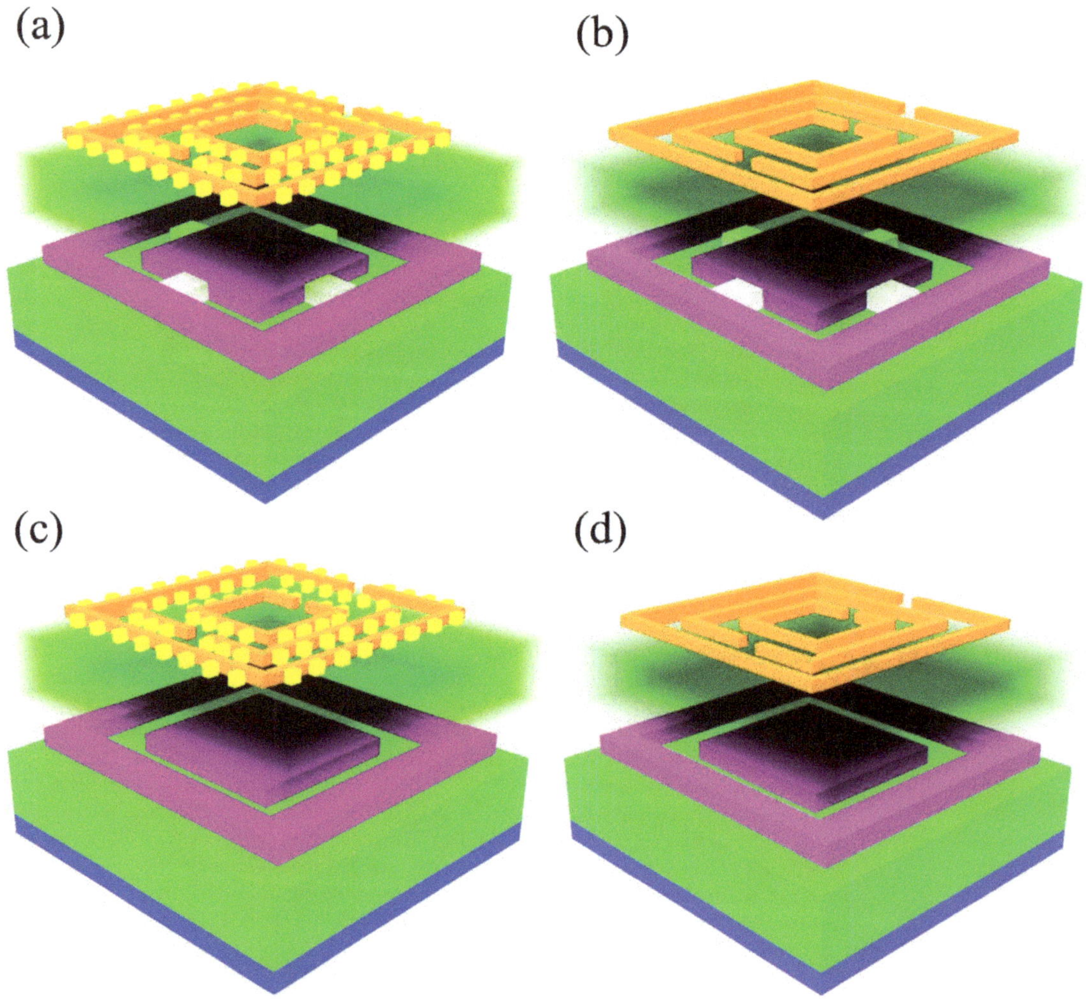

Figure 5. (**a**) The first design is the connected patch inner and patch outer (CPIO) and the SRR with the tooth (SWT) (**b**) Second design is the combined patch inner and patch outer (CPIO) and the SRR without a tooth (SWOT) (**c**) Third design is the disconnected patch inner and patch outer (DCPIO) and the SRR with the tooth (SWT) (**d**) Fourth design is the switch disconnected patch inner and patch outer (DCPIO) and the SRR without a tooth (SWOT).

Table 1. Different proposed structures.

Design	Antenna Design
1	Connected patch inner and patch outer (CPIO) and split ring resonator with tooth (SWT)
2	Connected patch inner and patch outer (CPIO) and split ring resonator without tooth (SWOT)
3	Disconnected patch inner and patch outer (DCPIO) and split ring resonator with tooth (SWT)
4	Disconnected patch inner and patch outer (DCPIO) and split ring resonator without tooth (SWOT)

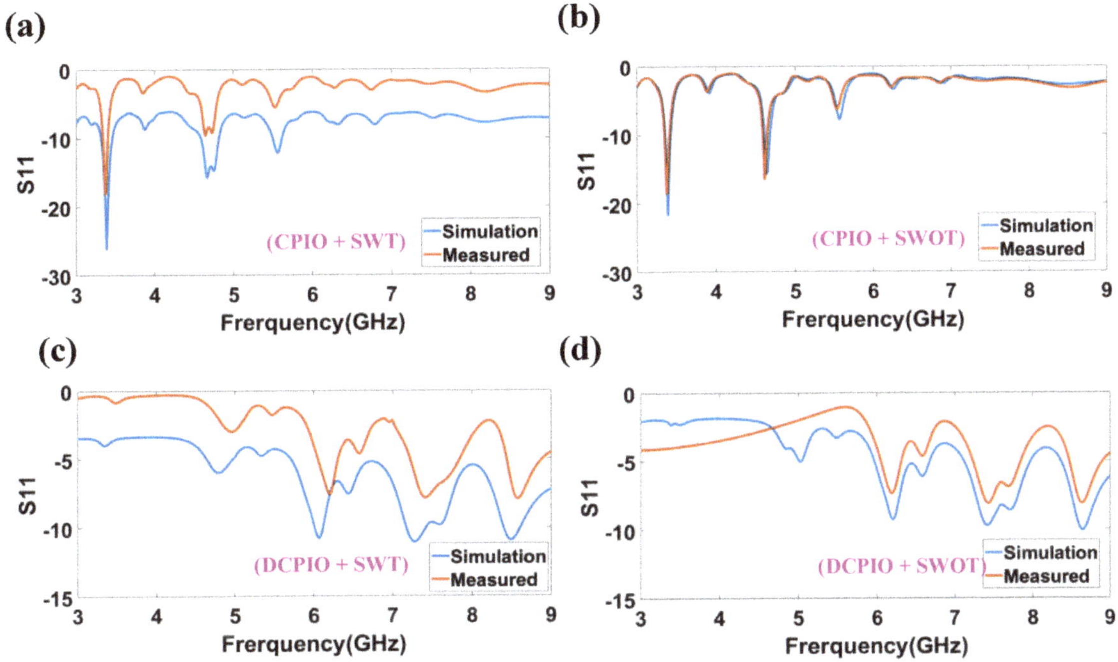

Figure 6. Measured and simulated reflectance response analysis using the FR-4 as substrate. The number of bands for the proposed four structures is 3, 2, 3 and 1. Minimum reflectance responses are, respectively, −26.11 dB,−21.52 dB, −11 dB and −10.10 dB. The resonance frequency is, respectively, 3.38 GHz, 3.39 GHz, 7.27 GHz and 8.64 GHz. (**a**) CPIO and SWT Design Structure. (**b**) CPIO and SWOT Design Structure. (**c**) DCPIO and SWT Structure. (**d**) DCPIO and SWOT Structure.

The voltage standing wave ratio (VSWR) represents how effectively the antenna power propagates through an antenna. Figure 8 (a) represents the VSWR response by choosing the substrate as FR4. The CPIO and SWT antenna structures provide a VSWR of 1.51 at 3.38 GHz. The CPIO and SWOT antenna structures provide a VSWR of 1.18 at 3.39 GHz. The DCPIO and SWT antenna structures provide a VSWR of 1.83 at 7.87 GHz. The DCPIO and SWOT antenna structures provide a VSWR of 2.21 at 8.64 GHz. It is observed that the connected interior and exterior patch (CPIO) design provides a better VSWR than the disconnected interior and exterior patch (DCIOP) design. Data are represented in Table 3. Figure 8b depicts the VSWR response by choosing the substrate as Rogers RT Duroid 5880. The CPIO and SWT antenna structures provide a VSWR of 1.05 at 6.49 GHz. The CPIO and SWOT antenna structures provide a VSWR of 1.03 at 6.52 GHz. The DCPIO and SWT antenna structures provide a VSWR of 2.33 at 8.83 GHz. The DCPIO and SWOT antenna structures provide a VSWR of 5.67 at 8.80 GHz. Data are represented in Table 2.

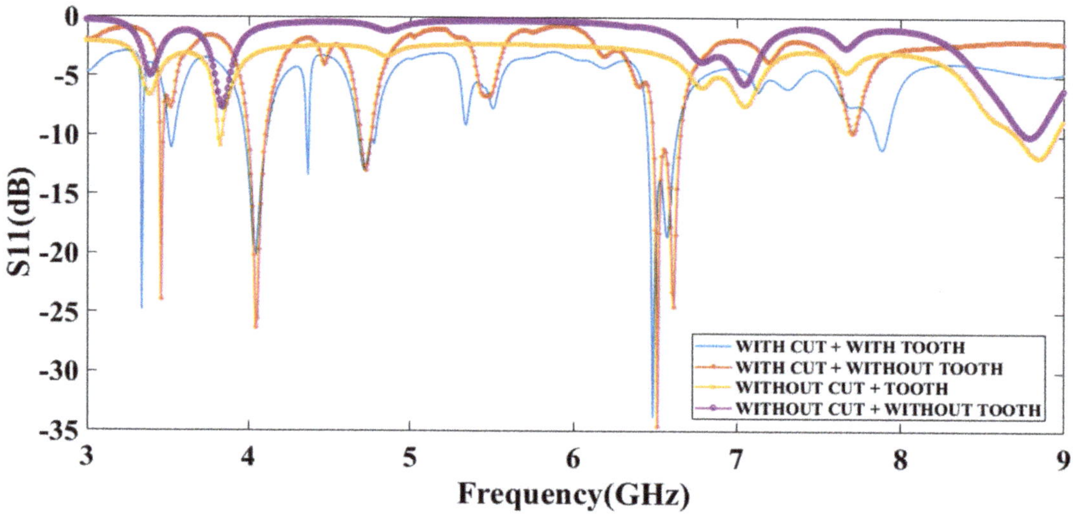

Figure 7. Simulated reflectance response analysis using Rogers RT Duroid as substrate. S_{11} plot for the proposed four antenna structures for the substrate Rogers RT Duroid 5880. The number of bands for the proposed four structures is 7, 6, 3 and 1. Minimum reflectance responses are, respectively, −33.79 dB, −34.54 dB, −13.79 dB and −10.05 dB. The resonance frequency is 6.49 GHz, 6.52 GHz, 8.83 GHz and 8.80 GHz, respectively.

Table 2. Reflectance plot (S_{11}), total bands, resonance frequency, VSWR, and bandwidth data representation for the proposed four antenna structures using a substrate of Rogers RT Duroid 5880.

Design	No of Bands	Reflectance Response (S_{11})	Resonance Frequency (GHz)	VSWR	Bandwidth (GHz)	Starting Point (GHz)	Ending Point (GHz)
CPIO and SWT	7.00	−24.78	3.33	1.15	0.02	3.32	3.34
		−11.11	3.52	2.00	0.03	3.51	3.54
		−20.11	4.04	1.28	0.13	3.98	4.11
		−13.64	4.37	1.73	0.01	4.36	4.37
		−13.00	4.72	1.78	0.11	4.68	4.79
		−33.79	6.49	1.05	0.18	6.46	6.64
		−11.17	7.89	2.06	0.07	7.85	7.92
CPIO and SWOT	6.00	−13.37	2.91	1.54	0.03	2.90	2.93
		−23.80	3.46	1.13	0.02	3.45	3.47
		−26.30	4.05	1.10	0.11	4.00	4.11
		−12.93	4.73	1.58	0.06	4.70	4.76
		−34.54	6.52	1.03	0.16	6.50	6.66
		−10.00	7.71	1.96	0.02	7.70	7.72
DCPIO and SWT	3.00	−12.89	3.83	3.00	0.06	3.80	3.86
		−10.00	7.06	6.77	0.02	7.05	7.07
		−13.79	8.83	2.33	0.51	8.54	9.05
DCPIO and SWOT	1.00	−10.05	8.80	5.67	0.07	8.76	8.83

Table 3. Reflectance plot (S_{11}), total bands, resonance frequency, VSWR, and bandwidth data representation for the proposed four antenna structures using a substrate of FR4.

Design	No of Bands	Reflectance Response (S_{11})	Resonance Frequency (GHz)	VSWR	Bandwidth (GHz)	Starting Point (GHz)	Ending Point (GHz)
CPIO and SWT	3.00	−26.11	3.38	1.51	0.13	3.32	3.45
		−15.70	4.66	5.21	0.22	4.59	4.81
		−12.30	5.56	8.00	0.11	5.50	5.61
CPIO and SWOT	2.00	−21.52	3.39	1.18	0.06	3.36	3.42
		−15.60	4.64	1.39	0.07	4.61	4.68
DCPIO and SWT	3.00	−10.74	6.06	1.99	0.10	6.01	6.11
		−11.00	7.27	1.83	0.24	7.17	7.41
		−10.89	8.49	1.87	0.20	8.40	8.60
DCPIO and SWOT	1.00	−10.10	8.64	2.21	0.02	8.63	8.65

An antenna's gain refers to how well it focuses or focuses the radio waves in a specific direction. The gain refers to an antenna's efficiency in transforming the electrical power into radio waves in one order. In contrast, the directivity refers to an antenna's capacity to concentrate the radiation in a single directional beam. Since many antennas and optical systems are only meant to emit electromagnetic waves in a particular direction or at a specific angle, the directivity is an essential metric to consider. Increases in directivity indicate that an antenna's emitted signal is being focused or directed more narrowly. Increasing the beam's directivity also increases its range. Since dBi is the standard unit of measurement for isotropic antennas, it is used to depict the antenna's performance, compared to the isotropic antenna [50]. Figure 9 represents the 3D radiation pattern plot for the FR-4 substrate (−180° to +180°). Figure 9a,b shows the maximum directivity of 5.59 dB and the normalized directivity of 81° (−39 to +42) achieved for the CPIO and SWT modes. Figure 9c,d shows the maximum directivity of 3.77 dB and the normalized directivity 93° (−39 to +54) achieved for the CPIO and SWOT models. Figure 9e,f show that the maximum directivity is 3.35 dB and the normalized directivity is 21° (−42 to −24), 18° (−8 to +10), 13° (28 to 41) achieved for the DCPIO and SWT modes. Figure 9g,h show the maximum directivity of 3.34 dB and the normalized directivity of 23° (−46 to −23), 17° (−7 to +10), 14° (+26 to +40) achieved for the DCPIO and SWOT models.

Figure 10 represents the directivity vs. the degree plot using Rogers RT Duriod 5880 as the substrate (−180° to +180°). Figure 10a,b show that the CPIO and SWT modes represent the maximum directivity of 6.86 dB and the normalized directivity of 24° (−16 to +8). The CPIO and SWOT models represent the maximum directivity of 6.84 dB and the normalized directivity of 15° (−49 to −35), 25° (−19 to +4), and 14° (+23 to +37). The DCPIO and SWT modes represent the maximum directivity of 4.47 dB and the normalized directivity of 32° (−43 to −11). The DCPIO and SWOT models represent the maximum directivity of 4.74 dB and the normalized directivity of 29° (−43 to −14).

The antenna's gain effectively represents the conversation of an applied electric signal into the electromagnetic waves. Figure 11 illustrates a total gain for the proposed four structures using the FR4, respectively, 4.54 dB, 3.51 dB, 2.39 dB and 2.38 dB. Figure 12 represents a total gain for the proposed four structures using Rogers RT Duroid 5880, respectively, 8.57 dB, 8.34 dB, 7.76 dB, and 7.47 dB. The comparison of the total gain for the proposed four antenna structures using the different substrate materials are represented in Table 4. The comparison of proposed design with another multiband design is represented in the Table 5.

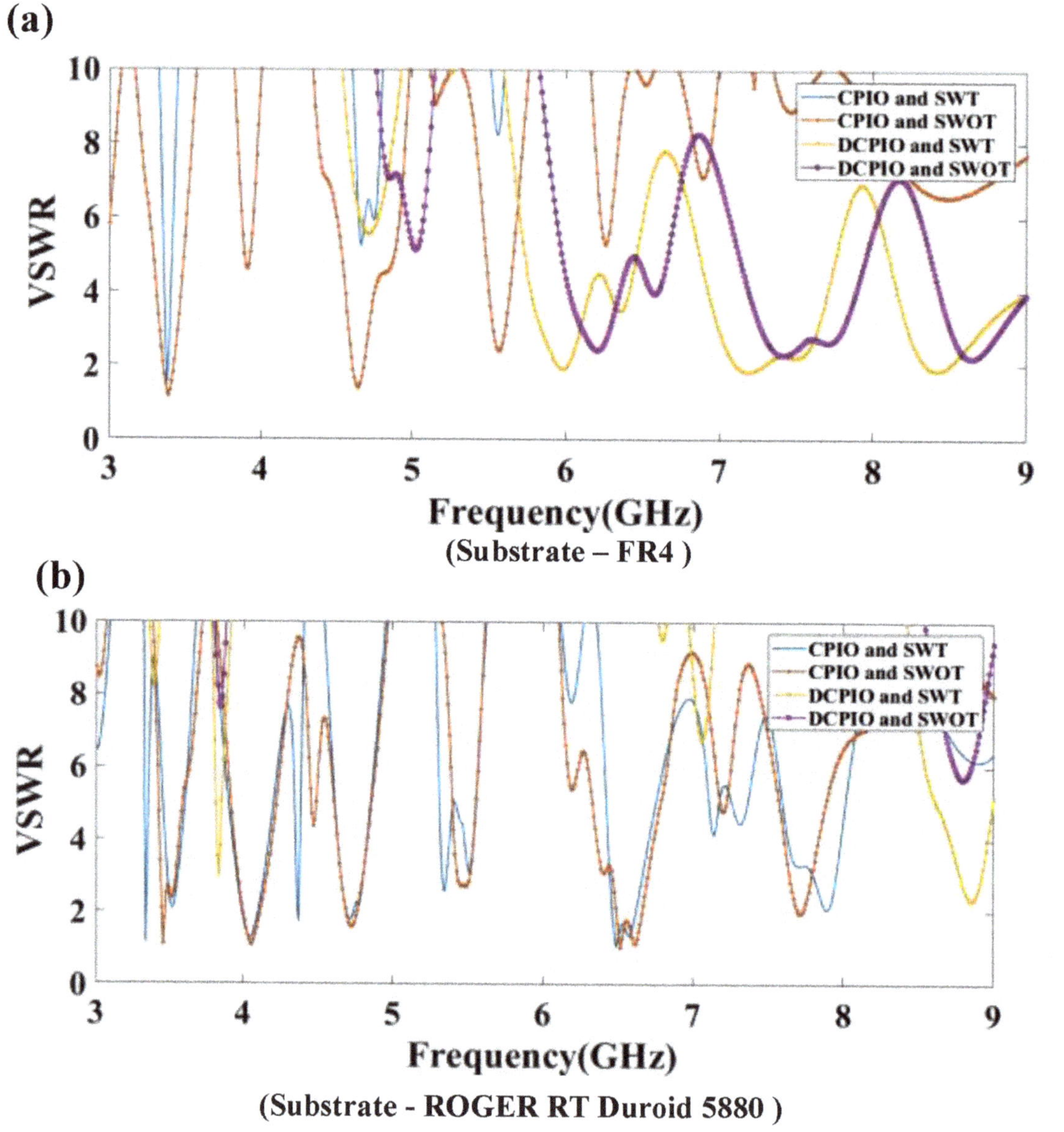

Figure 8. (**a**) The minimum value of the VSWR for the proposed four antenna structures by selecting the substrate as FR4, is 1.51 at 3.38 GHz, 1.18 at 3.39 GHz, 1.83 at 7.27 GHz, respectively, and 2.21 at 8.64 GHz. (**b**) The minimum value of the VSWR for the proposed four antenna structures by selecting the substrate as Rogers RT Duroid 5880 are 1.05 at 6.49 GHz, 1.03 at 6.52 GHz, 2.33 at 8.83 GHz and 5.67 at 8.80 GHz.

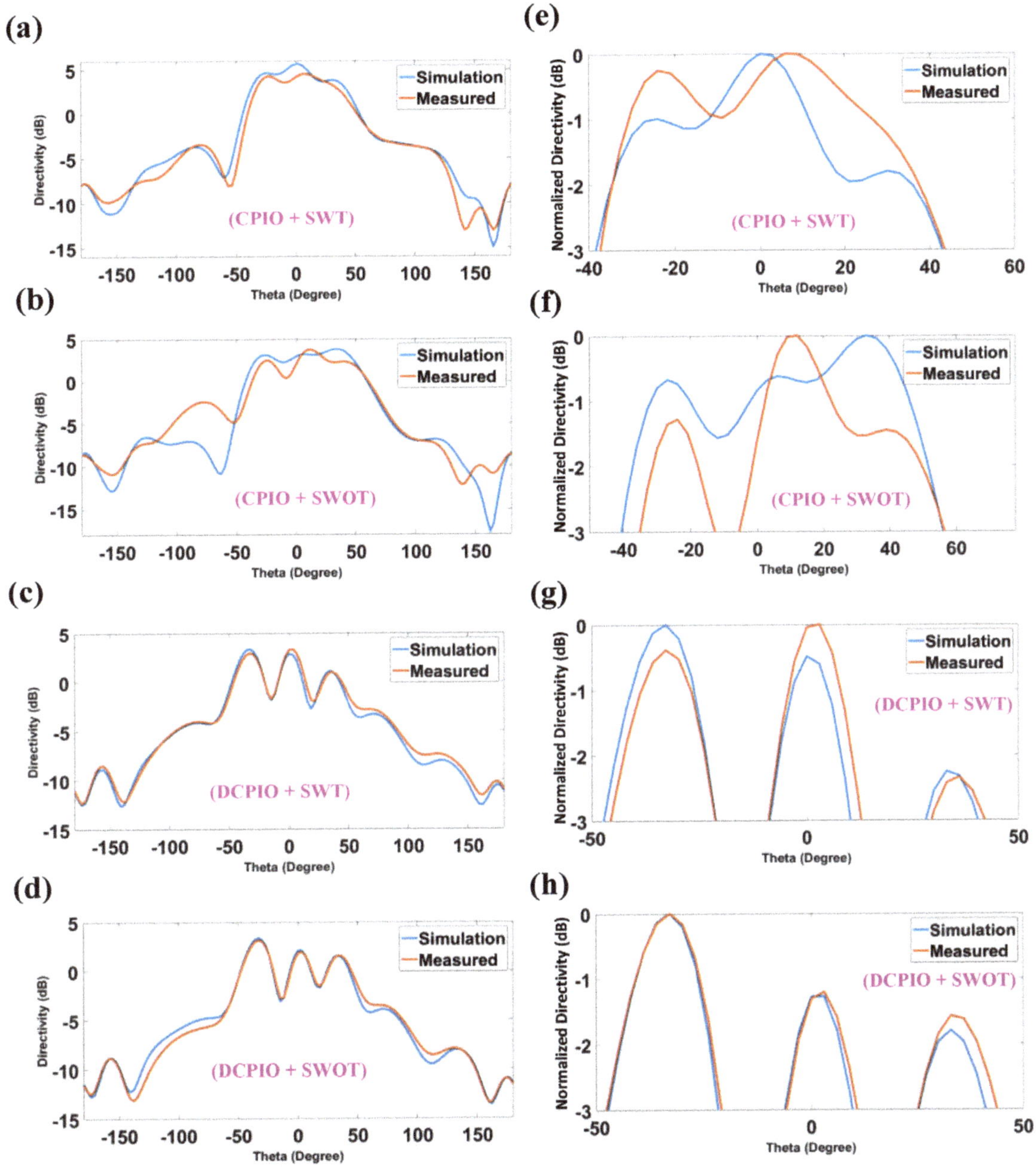

Figure 9. Maximum directivity and the normalized directivity for all of the modes are, respectively, 5.59 dB with 81°, 3.77 dB with 93°, 3.35 dB with 21°, 18°, 13° and 3.34 dB with 23°, 17°, 14°.

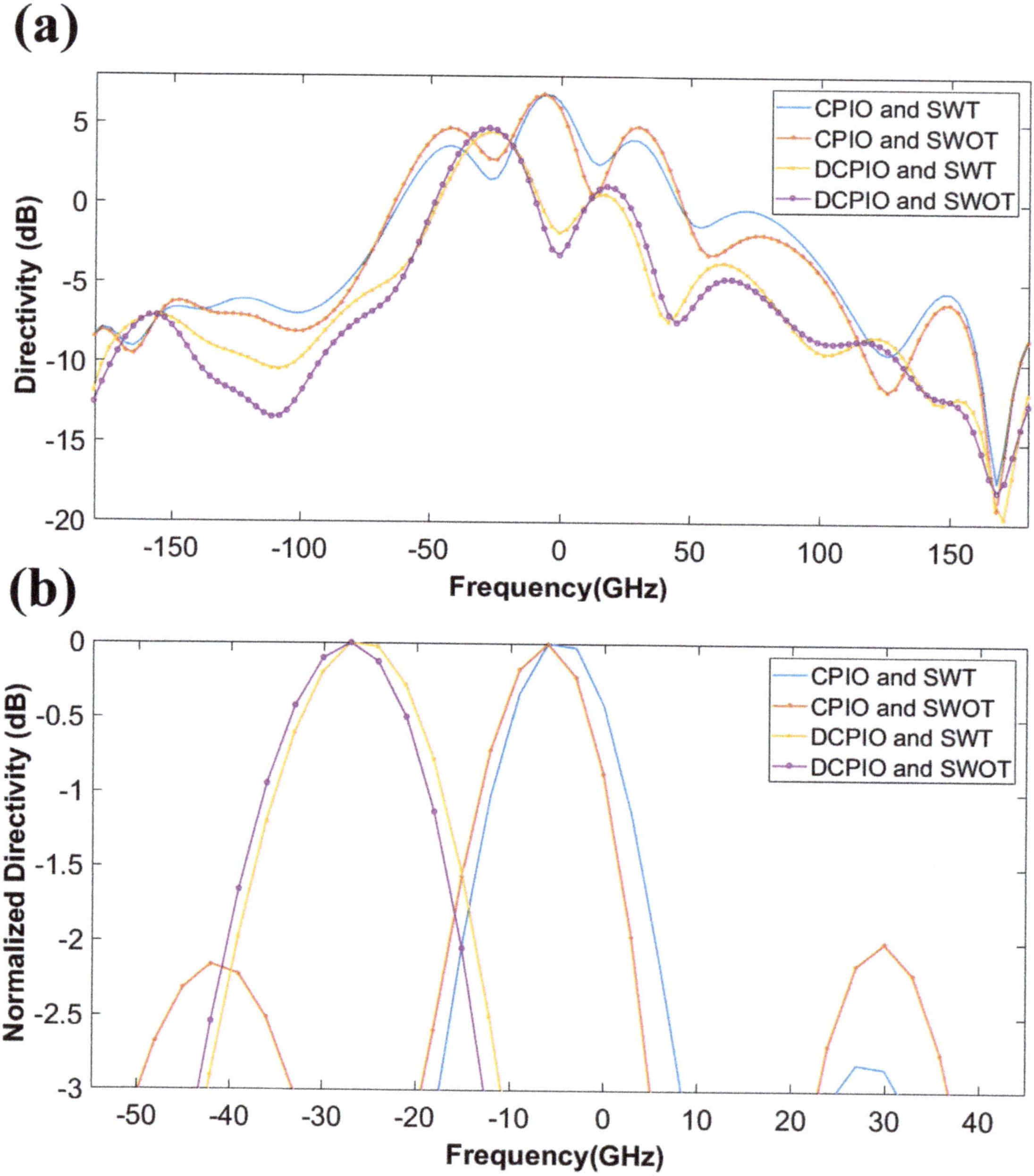

Figure 10. Directivity plot using the Rogers RT Duroid 5880 substrate for the range of −180° to +180°. The maximum directivity and the normalized directivity for all of the modes are 6.86 dB with 24° and 7°, 6.84 dB with 15°, 25°, 14°, and 4.47 dB with 32° and 4.74 dB with 29°.

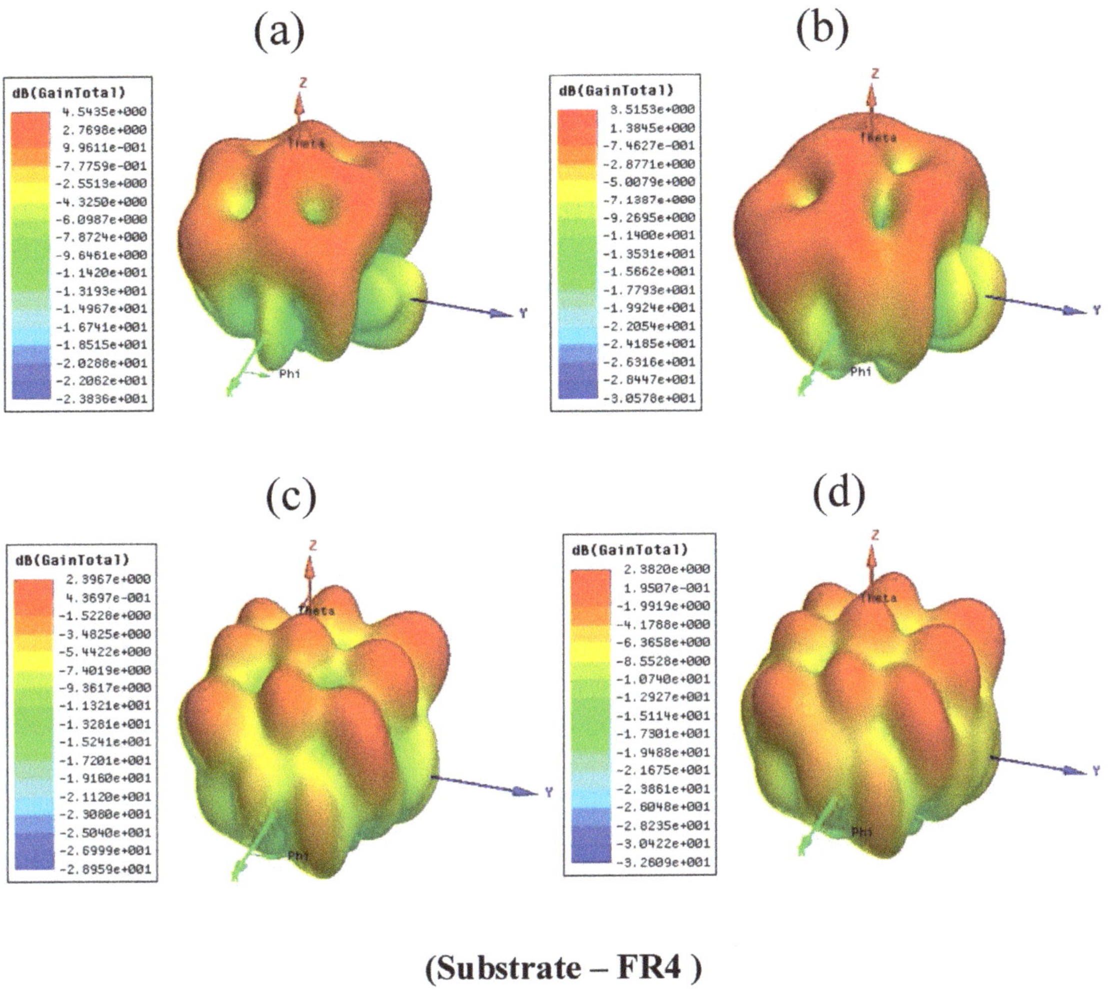

Figure 11. Total gain for the presented four antenna structures by selecting the substrate FR4. (**a**) 4.54 dB for the CPIO and SWT (**b**) 3.51 dB for the CPIO and SWOT (**c**) 2.39 dB for the DCPIO and SWT (**d**) 2.38 dB for the DCPIO and SWOT.

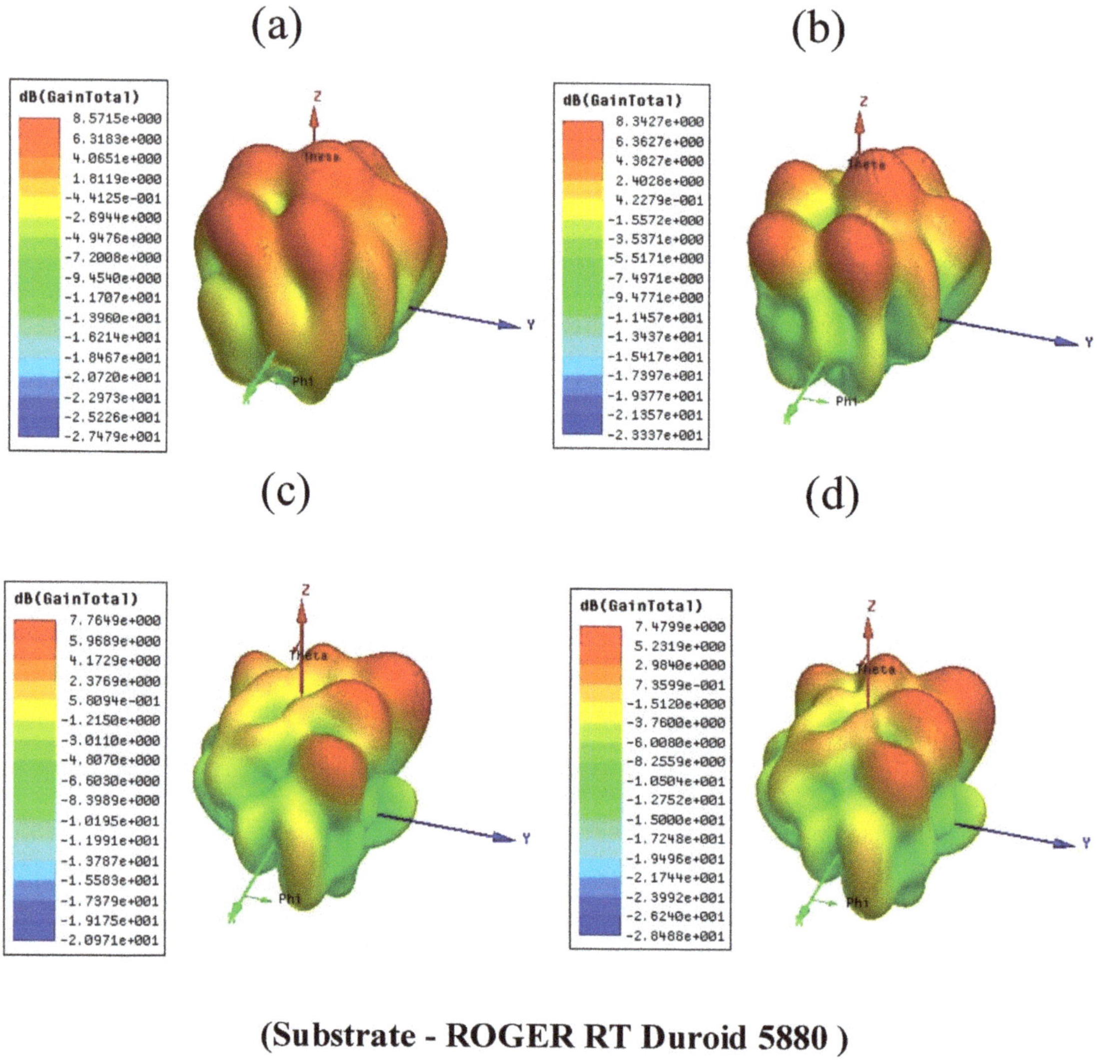

Figure 12. Total gain for the presented four antenna structures by selecting the substrate Rogers RT Duroid 5880. (**a**) 8.57 dB for the CPIO and SWT (**b**) 8.34 dB for the CPIO and SWOT (**c**) 7.76 dB for the DCPIO and SWT (**d**) 7.7 dB for the DCPIO and SWOT.

Table 4. Total gain for the two different substrates, Rogers RT Duroid 5880 and FR4.

Design	Substrate	
	Rogers RT Duroid 5880	**FR4**
CPIO and SWT	8.57	4.54
CPIO and SWOT	8.34	3.51
DCPIO and SWT	7.76	2.39
DCPIO and SWOT	7.47	2.38

Table 5. Comparison of the proposed structure with another multiband design.

References	No of Bands	Resonating Frequency (GHz)	Minimum Reflectance Response (S_{11})	Peak Gain (dB)
[44]	5	4.4, 5, 5.8, 8.05	−17, −31, −13, −17	-
	4	2.9, 5.1, 5.95, 6.55, 8.3	−25, −13, −22, −25, −21.5	-
	5	3.2, 5.5, 6, 6.6, 8.3	−10, −17, −22, −12, −10	-
[9]	3	3.4, 5.7	−20, −39	-
[35]	5	5.3, 7.5, 9.8, 14.9, 19	−13, −12, −10, −23, −21	8.5
[6]	2	5.7, 10.3	−22, −21	6.38
[28]	3	4, 4.8, 9	−26, −16.5, −35	3.24
Proposed CPIO and SWT structure	7	3.33, 3.52, 4.04, 4.37, 4.72, 6.49, 7.89	−24.78, −11.11, −20.11, −13.64, −13, −33.79, −11.17	8.57

4. Conclusions

It is concluded that the tooth-added metamaterial superstrate structure with the connected interior and exterior patch provides the multiband operation with a healthy gain. The measured and fabricated results are compared for verification. The system's performance is analyzed by varying the substrate material, inserting and removing the tooth in the split-ring resonator, and connecting and disconnecting the interior and exterior patch. The results are compared, in terms of many bands, S_{11}, the voltage standing wave ratio and the bandwidth for the proposed four structures, by changing the substrate material Rogers RT Duroid 5880 and FR4. The results are compared with earlier published work. The presented design represents seven frequency bands of operation, a gain of 8.57 dB, and a maximum reflectance response of −33.79 dB. The presented structure is appropriate for numerous wireless communication applications, such as radar surveillance, satellite communication and weather monitoring.

Author Contributions: Conceptualization, K.A. and S.K.P.; methodology, M.A. and S.L.; software, S.L. and K.A.; validation, S.L., K.A., S.K.P. and M.G.D.; formal analysis, A.A. and M.A.; investigation, A.A. and M.G.D.; resources, K.A., A.A. and M.A.; writing—original draft preparation, All authors.; writing—review and editing, A.A. and S.K.P.; visualization, K.A. and M.A.; supervision, A.A. and S.K.P.; project administration. All authors have read and agreed to the published version of the manuscript.

Funding: This research received no external funding.

Data Availability Statement: The data will be made available at a reasonable request to the corresponding author.

Conflicts of Interest: The authors declare no conflict of interest.

References

1. Liang, J.; Yang, H.Y.D. Microstrip Patch Antennas on Tunable Electromagnetic Band-Gap Substrates. *IEEE Trans. Antennas Propag.* **2009**, *57*, 1612–1617. [CrossRef]
2. Patel, S.K.; Sorathiya, V.; Sbeah, Z.; Lavadiya, S.; Nguyen, T.K.; Dhasarathan, V. Graphene-Based Tunable Infrared Multi Band Absorber. *Opt. Commun.* **2020**, *474*, 126109. [CrossRef]
3. Divya, C.; Koushick, V. Design and Implementation of Slotted Metamaterial Stacked Microstrip Patch Antenna for Broadband Applications. *J. Phys. Conf. Ser.* **2020**, *1432*, 012067. [CrossRef]
4. Hussain, N.; Jeong, M.J.; Abbas, A.; Kim, T.J.; Kim, N. A Metasurface-Based Low-Profile Wideband Circularly Polarized Patch Antenna for 5G Millimeter-Wave Systems. *IEEE Access* **2020**, *8*, 22127–22135. [CrossRef]
5. Patel, S.K.; Shah, K.H.; Kosta, Y.P. Frequency-Reconfigurable and High-Gain Metamaterial Microstrip-Radiating Structure. *Waves Random Complex Media* **2019**, *29*, 523–539. [CrossRef]
6. Rajak, N.; Chattoraj, N.; Mark, R. Metamaterial Cell Inspired High Gain Multiband Antenna for Wireless Applications. *AEU - Int. J. Electron. Commun.* **2019**, *109*, 23–30. [CrossRef]
7. Nguyen, T.K.; Patel, S.K.; Lavadiya, S.; Parmar, J.; Bui, C.D. Design and Fabrication of Multiband Reconfigurable Copper and Liquid Multiple Complementary Split-Ring Resonator Based Patch Antenna. *Waves Random Complex Media* **2022**, 1–24. [CrossRef]

8. Piper, B.R.; Bialkowski, M.E. Modelling the Effect of 3D Conformity on Single and Multi-Band Patch Antennas. In Proceedings of the IEEE Antennas and Propagation Society, AP-S International Symposium (Digest), Columbus, OH, USA, 22–27 June 2003; Volume 3, pp. 116–119.
9. Yu, K.; Li, Y.; Wang, Y. Multi-Band Metamaterial-Based Microstrip Antenna for WLAN and WiMAX Applications. In Proceedings of the 2017 International Applied Computational Electromagnetics Society Symposium, ACES, Florence, Italy, 26–30 March 2017; pp. 1–2.
10. Yoon, C.; Choi, S.H.; Lee, H.C.; Park, H.D. Small Microstrip Patch Antennas with Short-Pin Using a Dual-Band Operation. *Microw. Opt. Technol. Lett.* **2008**, *50*, 367–371. [CrossRef]
11. Ding, Y.; Li, M.Q.; Chang, H.X.; Qin, K. A Dual-Band High Gain Antenna Based on Split Ring Resonators and Corrugated Plate. *Prog. Electromagn. Res. Lett.* **2014**, *44*, 87–92. [CrossRef]
12. Antar, Y.M.M. Microstrip Antenna Design Handbook. *IEEE Antennas Propag. Mag.* **2003**, *45*, 86. [CrossRef]
13. Patel, S.K.; Lavadiya, S.; Kosta, Y.P.; Kosta, M.; Nguyen, T.K.; Dhasarathan, V. Numerical Investigation of Liquid Metamaterial-Based Superstrate Microstrip Radiating Structure. *Phys. B Condens. Matter* **2020**, *585*, 412095. [CrossRef]
14. Lavadiya, S.P.; Sorathiya, V.; Kanzariya, S.; Chavda, B.; Naweed, A.; Faragallah, O.S.; Eid, M.M.A.; Rashed, A.N.Z. Low Profile Multiband Microstrip Patch Antenna with Frequency Reconfigurable Feature Using PIN Diode for S, C, X, and Ku Band Applications. *Int. J. Commun. Syst.* **2022**, *35*, e5141. [CrossRef]
15. Patel, S.K.; Kosta, Y.P. Meandered Multiband Metamaterial Square Microstrip Patch Antenna Design. *Waves Random Complex Media* **2012**, *22*, 475–487. [CrossRef]
16. Godaymi Al-Tumah, W.A.; Shaaban, R.M.; Ahmed, Z.A. A Modified E-Shaped Microstrip Patch Antenna for Dual Band in x-and Ku-Bands Applications. *J. Phys. Conf. Ser.* **2019**, *1234*, 012028. [CrossRef]
17. Yang, F.; Rahmat-Samii, Y. A Compact Dual Band Circularly Polarized Antenna Design for Mars Rover Mission. In Proceedings of the IEEE Antennas and Propagation Society, AP-S International Symposium (Digest), Columbus, OH, USA, 22–27 June 2003; Volume 3, pp. 858–861.
18. Ismail, M.K.H.; Esa, M. Low Profile Printed Antenna with a Pair of Step-Loading for Dual-Frequency Operation. In Proceedings of the 2003 Asia-Pacific Conference on Applied Electromagnetics, APACE, Shah Alam, Malayasia, 12–14 August 2003; pp. 54–57.
19. Patel, S.K.; Kosta, Y.P.; Charola, S. Liquid Metamaterial Based Radome Design. *Microw. Opt. Technol. Lett.* **2018**, *60*, 2303–2309. [CrossRef]
20. Viswanadha, K.; Raghava, N.S. Design and Analysis of a Multi-Band Flower Shaped Patch Antenna for WLAN/WiMAX/ISM Band Applications. *Wirel. Pers. Commun.* **2020**, *112*, 863–887. [CrossRef]
21. Lier, E.; Jakobsen, K.R. Rectangular Microstrip Patch Antennas with Infinite and Finite Ground Plane Dimensions. *IEEE Trans. Antennas Propag.* **1983**, *31*, 978–984. [CrossRef]
22. Thai, T.T.; DeJean, G.R.; Tentzeris, M.M. Design and Development of a Novel Compact Soft-Surface Structure for the Front-to-Back Ratio Improvement and Size Reduction of a Microstrip Yagi Array Antenna. *IEEE Antennas Wirel. Propag. Lett.* **2008**, *7*, 369–373. [CrossRef]
23. Lavadiya, S.P.; Patel, S.K.; Ahmed, K.; Taya, S.A.; Das, S.; Babu, K.V. Design and Fabrication of Flexible and Frequency Reconfigurable Antenna Loaded with Copper, Distilled Water and Seawater Metamaterial Superstrate for IoT Applications. *Int. J. RF Microw. Comput. Eng.* **2022**, *32*, e23481. [CrossRef]
24. Rajesh, G.S.; Kishore, K.V.; Kumar, V. Multiband Microstrip Patch Antenna Design Using Metamaterial for Airborne SAR System. In Proceedings of the 2015 IEEE International Conference on Signal Processing, Informatics, Communication and Energy Systems, SPICES, Calicut, India, 19–21 February 2015; pp. 1–3.
25. Sharma, N.; Vyas, K.; Srivastava, R. Analysis and Design of Microstrip Patch Antenna with Two Different Metamaterial Unit Cells. In *Flexible Electronics for Electric Vehicles*; Springer: Singapore, 2023; pp. 577–588.
26. Singh, A.K.; Inamdar, K. Fractal CSRR Metamaterial-Based Wearable Antenna for IoT Application. In *Emerging Technology Trends in Electronics, Communication and Networking*; Springer: Singapore, 2023; pp. 105–117.
27. Tiwari, D.; Ansari, J.A.; Saroj, A.K.; Kumar, M. Analysis of a Miniaturized Hexagonal Sierpinski Gasket Fractal Microstrip Antenna for Modern Wireless Communications. *AEU - Int. J. Electron. Commun.* **2020**, *123*, 153288. [CrossRef]
28. Patel, S.K.; Argyropoulos, C.; Kosta, Y.P. Broadband Compact Microstrip Patch Antenna Design Loaded by Multiple Split Ring Resonator Superstrate and Substrate. *Waves Random Complex Media* **2017**, *27*, 92–102. [CrossRef]
29. Sumathi, K.; Lavadiya, S.; Yin, P.Z.; Parmar, J.; Patel, S.K. High Gain Multiband and Frequency Reconfigurable Metamaterial Superstrate Microstrip Patch Antenna for C/X/Ku-Band Wireless Network Applications. *Wirel. Netw.* **2021**, *27*, 2131–2146. [CrossRef]
30. Methfesse, S.; Schmidt, L.P. Design of a Balanced-Fed Patch-Excited Horn Antenna at Millimeter-Wave Frequencies. In Proceedings of the EuCAP 2010—The 4th European Conference on Antennas and Propagation, Barcelona, Spain, 12–16 April 2010; pp. 1–4.
31. Zhu, H.; Cheung, S.W.; Yuk, T.I.P. Enhancing Antenna Boresight Gain Using a Small Metasurface Lens: Reduction in Half-Power Beamwidth. *IEEE Antennas Propag. Mag.* **2016**, *58*, 35–44. [CrossRef]
32. Latif, S.I.; Shafai, L.; Shafai, C. Gain and Efficiency Enhancement of Compact and Miniaturised Microstrip Antennas Using Multi-Layered Laminated Conductors. *IET Microw. Antennas Propag.* **2011**, *5*, 402–411. [CrossRef]
33. Lavadiya, S.P.; Patel, S.K.; Rayisyan, M. High Gain and Frequency Reconfigurable Copper and Liquid Metamaterial Tooth Based Microstrip Patch Antenna. *AEU - Int. J. Electron. Commun.* **2021**, *137*, 153799. [CrossRef]

34. Slyusar, V.I. Metamaterials on Antenna Solutions. In Proceedings of the International Conference on Antenna Theory and Techniques, Lviv, Ukraine, 6–9 October 2009; pp. 19–24.
35. Madhav, B.T.P.; Manjeera, M.; Navya, M.S.; Sharada Devi, D.; Sumanth, V. Novel Metamaterial Loaded Multiband Patch Antenna. *Indian J. Sci. Technol.* **2016**, *9*, 1–9. [CrossRef]
36. Palandoken, M. Artificial Materials Based Microstrip Antenna Design. In *Microstrip Antennas;* InTech: Rijeka, Croatia, 2011.
37. Jaydeep, S.; Sunil, L. An Investigation on Recent Trends in Metamaterial Types and Its Applications. *i-Manager's J. Mater. Sci.* **2018**, *5*, 55. [CrossRef]
38. Lavadiya, S.; Sorathiya, V.; Das, S. Simulation and Fabrication of High Gain Diffracted Ground-Based Metamaterial Microstrip Patch Antenna for C Band. In *Lecture Notes in Electrical Engineering;* Springer: Berlin/Heidelberg, Germany, 2022; Volume 851, pp. 403–412. ISBN 9789811691539.
39. Jackson, D.R.; Williams, J.T.; Bhattacharyya, A.K.; Smith, R.L.; Buchheit, S.J.; Long, S.A. Microstrip Patch Designs That Do Not Excite Surface Waves. *IEEE Trans. Antennas Propag.* **1993**, *41*, 1026–1037. [CrossRef]
40. Pirhadi, A.; Hakkak, M.; Keshmiri, F. Using Electromagnetic Bandgap Superstrate to Enhance the Bandwidth of Probe-Fed Microstrip Antenna. *Prog. Electromagn. Res.* **2006**, *61*, 215–230. [CrossRef]
41. Fangming, Z.; Qingchun, L.; Jun, H. A Directive Patch Antenna with a Metamaterial Cover. In Proceedings of the Asia-Pacific Microwave Conference Proceedings, APMC, Suzhou, China, 4–7 December 2005; Volume 3, pp. 1–3.
42. Mosallaei, H.; Sarabandi, K. Antenna Miniaturization and Bandwidth Enhancement Using a Reactive Impedance Substrate. *IEEE Trans. Antennas Propag.* **2004**, *52*, 2403–2414. [CrossRef]
43. Mosallaei, H.; Sarabandi, K. A Novel Artificial Reactive Impedance Surface for Miniaturized Wideband Planar Antenna Design: Concept and Characterization. In Proceedings of the IEEE Antennas and Propagation Society, AP-S International Symposium (Digest), Columbus, OH, USA, 22–27 June 2003; Volume 2, pp. 403–406.
44. Patel, S.K.; Lavadiya, S.P.; Parmar, J.; Ahmed, K.; Taya, S.A.; Das, S. Low-Cost, Multiband, High Gain and Reconfigurable Microstrip Radiating Structure Using PIN Diode for 5G/Wi-MAX/WLAN Applications. *Phys. B Condens. Matter* **2022**, *639*, 413972. [CrossRef]
45. Patel, S.K.; Lavadiya, S.P.; Parmar, J.; Das, S.; Ahmed, K.; Taya, S.A. Low-Cost, Compact, and Reconfigurable Antennas Using Complementary Split-Ring Resonator Metasurface for next-Generation Communication Systems. *Int. J. Microw. Wirel. Technol.* **2022**, 1–11. [CrossRef]
46. D.P. CRC Handbook of Chemistry and Physics: Editor-in-chief D.R. Lide; CRC Press, Boca Raton, FL, USA, 71st edn, 1990–1991, pp. 2324. *J. Mol. Struct.* **1992**, *268*, 320. [CrossRef]
47. Lavadiya, S.P.; Sorathiya, V.; Kanzariya, S.; Chavda, B.; Faragallah, O.S.; Eid, M.M.A.; Rashed, A.N.Z. Design and Verification of Novel Low-Profile Miniaturized Pattern and Frequency Tunable Microstrip Patch Antenna Using Two PIN Diodes. *Braz. J. Phys.* **2021**, *51*, 1303–1313. [CrossRef]
48. Patel, S.K.; Kosta, Y. Investigation on Radiation Improvement of Corner Truncated Triband Square Microstrip Patch Antenna with Double Negative Material. *J. Electromagn. Waves Appl.* **2013**, *27*, 819–833. [CrossRef]
49. Balanis, C.A. *Antenna Theory: Analysis and Design;* John Wiley and Sons: Hoboken, NJ, USA, 2016.
50. Stutzman, W.L.; Thiele, G.A. *Antenna Theory and Design;* John Wiley & Sons: Hoboken, NJ, USA, 2012.

Disclaimer/Publisher's Note: The statements, opinions and data contained in all publications are solely those of the individual author(s) and contributor(s) and not of MDPI and/or the editor(s). MDPI and/or the editor(s) disclaim responsibility for any injury to people or property resulting from any ideas, methods, instructions or products referred to in the content.

micromachines

MDPI

Article

Dual-Band MIMO Antenna with Enhanced Isolation for 5G NR Application

Shuqi Xi, Jing Cai, Lingrong Shen, Qiangjuan Li and Gui Liu *

College of Electrical and Electronic Engineering, Wenzhou University, Wenzhou 325035, China
* Correspondence: gliu@wzu.edu.cn

Abstract: A two-port multiple-input and multiple-output (MIMO) antenna with dual-band characteristics operating at the fifth-generation (5G) new radio (NR) sub-6 GHz n7/n38/n41/n79 bands is proposed. The proposed MIMO antenna is composed of two symmetric antenna elements and a defected ground plane. The antenna element consists of an incomplete circular patch with two L-shaped branches. By applying the defected ground structure and the slotted stub, the current distribution on the ground plane is changed to reduce the mutual coupling between the antenna elements. The measured −10 dB reflection coefficients cover 2.34–2.71 GHz and 3.72–5.10 GHz, while the measured isolation is larger than 20 dB at the whole operating frequency band. The paper has investigated different performance parameters in terms of the envelope correction coefficient (ECC), diversity gain (DG), radiation patterns, antenna gain, and efficiency. The proposed MIMO antenna is suitable for 5G applications.

Keywords: dual-band; 5G NR; high isolation; MIMO antenna

1. Introduction

Nowadays, the fifth generation (5G) communication system, when it is compared with traditional communication systems, has the advantages of high speed, wide frequency bandwidth, low power consumption, and high reliability. Meanwhile, the multiple-input and multiple-output (MIMO) antenna system has attracted extensive attention due to it having lower multipath fading effects, higher channel capacity, and an increased transmission rate than those of the single antenna communication system. However, there are still many challenges in designing high-isolation and compact-size MIMO antennas, such as the mutual coupling between the antenna elements [1].

Therefore, it is a critical problem in the design process of the MIMO antenna to reduce mutual coupling. Many scholars have presented various decoupling techniques to reduce mutual coupling between antennas [2–7]. In [2], an ultrawideband (UWB) MIMO antenna achieved high isolation values that are larger than 20 dB for the whole operating band by slotted stubs. A coplanar waveguide-fed MIMO antenna with high isolation values was proposed by adding a double Y-shaped branch [3]. In [4], by introducing a creative un-protruded multi-slot (UPMS) isolating element, a four-port MIMO antenna array was proposed to reduce the mutual coupling between the antenna elements further. In [5], each antenna element was placed orthogonally to achieve the good isolation of 45 dB at the band of 36.83–40.0 GHz. It is reported in [6] that a combination of a defected ground structure (DGS) and electromagnetic band gaps (EBG) can be adopted to ease mutual coupling. In [8], the mutual coupling of a four-port MIMO antenna has been reduced by the orthogonal orientation of radiating elements rather than any decoupling structures.

Furthermore, it is also challenging to design a multi-band MIMO antenna to meet substantial wireless communication bands because the same isolation method may not work at different frequency bands. Recently, several dual-band MIMO antennas have been proposed [9–20]. In [9], a dual-notched four-element MIMO antenna with dual-band

Citation: Xi, S.; Cai, J.; Shen, L.; Li, Q.; Liu, G. Dual-Band MIMO Antenna with Enhanced Isolation for 5G NR Application. *Micromachines* **2023**, *14*, 95. https://doi.org/10.3390/mi14010095

Academic Editors: Trushit Upadhyaya and Hari Shankar Singh

Received: 25 November 2022
Revised: 14 December 2022
Accepted: 21 December 2022
Published: 30 December 2022

Copyright: © 2022 by the authors. Licensee MDPI, Basel, Switzerland. This article is an open access article distributed under the terms and conditions of the Creative Commons Attribution (CC BY) license (https://creativecommons.org/licenses/by/4.0/).

characteristics has been proposed by introducing gap sleeves and an H slot. It covers the frequency bands of 3.3–4.1 GHz and 8.2–8.6 GHz. A dual-band MIMO antenna was proposed in [10] by implementing an L-shaped feeding strip, a parasitic rectangle strip, and a modified Z-shaped radiating strip. In [13], an eight-element MIMO antenna was designed to meet the requirement of the 5G mobile terminals. A quad-element MIMO antenna with E-shaped and G-shaped stubs was presented to achieve two resonances at 1.5 GHz and 2.45 GHz, respectively in [18]. In [20], a MIMO antenna array was realized which can cover the 5G NR Bands n77/n78/n79 and WLAN 5 GHz band.

In this paper, a two-port MIMO antenna with high isolation and dual-band characteristics is presented. The antenna consists of two identical three-quarter circular patches which are placed symmetrically on the top of the substrate, with the size of $41 \times 30 \times 1.59$ mm^3. The antenna generates two frequency resonances by introducing two L-shaped branches and two step-shaped gaps in the radiators. The working bands can cover n7 (2.5–2.69 GHz), n38 (2.57–2.62 GHz), n41 (2.496–2.690 GHz), and n79 bands (4.4–5 GHz). The measured isolation is larger than 20 dB. Other MIMO antenna parameters are presented to analyze the performance, such as the S-parameters (reflection and transmission coefficients), radiation patterns, ECC, DG, and efficiencies.

2. Antenna Design and Analysis

2.1. Antenna Geometry

The configuration and prototype of the designed dual-band antenna are shown in Figure 1. The radiator of the antenna element consists of an L-shaped strip and a defective circular radiator composed of a semicircle with step-shaped strips. The radiator is fed by a 50 Ω micro-strip feedline. Two identical antenna elements are printed on a 1.59 mm thick FR4 substrate with relative permittivity of 4.4 and loss tangent of 0.02. A cambered ground plane is printed on the bottom of the substrate. High isolation is achieved by etching a C-shaped slot, a rectangular slot, and a rectangular slit. Through simulation and optimization, the detailed parameters are shown in Table 1.

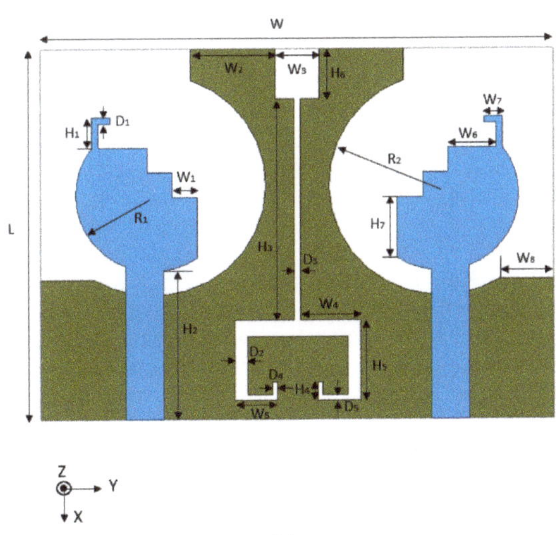

(a)

Figure 1. *Cont.*

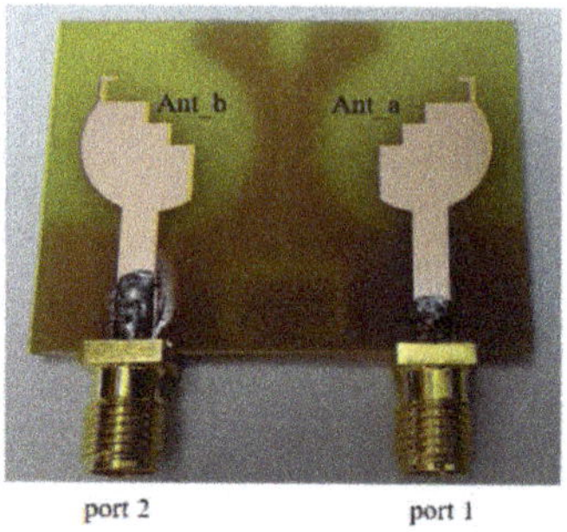

(**b**)

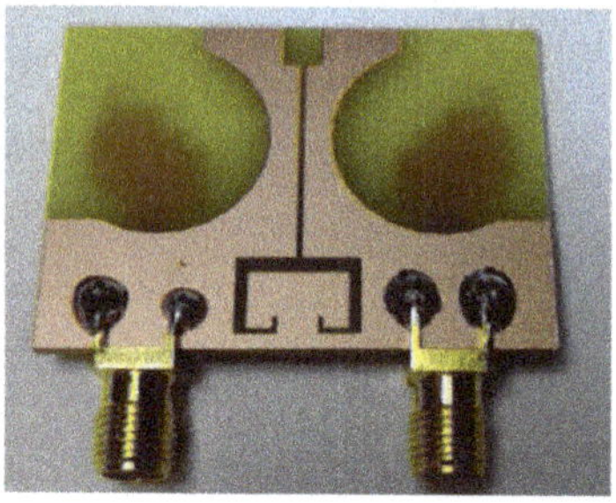

(**c**)

Figure 1. The geometry and prototype of the proposed design: (**a**) perspective view, (**b**) top view, and (**c**) bottom view.

Table 1. Parameters of the presented antenna (unit: mm).

Parameter	W_1	W_2	W_3	W_4	W_5	W_6
Value	2	6.75	3.5	4.75	3.4	3.9
Parameter	W_7	W_8	H_1	H_2	H_3	H_4
Value	1.5	4.34	2.5	12.05	18	1.5
Parameter	H_5	H_6	D_1	D_2	D_3	D_4
Value	6.5	4	0.5	1	0.5	0.4
Parameter	D_5	R_1	R_2	W	L	
Value	0.5	6.2	9	41	30	

2.2. Design Process

Figure 2 presents the detailed process for the proposed antenna to study the performance of the 5G dual-band monopole antenna. Because two antenna elements are placed symmetrically to share the ground plane, S_{11} and S_{21} are the same as S_{22} and S_{12}, respectively. For simplicity, we only discuss S_{11} and S_{21}. The simulated S-parameters are plotted in Figure 3. In Antenna 1, a three-quarters circular radiation patch with the circular slotted ground plane is designed. Antenna 1 can achieve a −10 dB impedance bandwidth of 500 MHz (2–2.5 GHz) and 1880 MHz (4.05–5.93 GHz). However, the isolation between the radiating elements is non-ideal over the operation bands, which is illustrated in Figure 3b. In Antenna 2, an L-shaped branch is designed to realize a lower frequency shift (right) and a higher frequency shift (left), which can be seen in Figure 3a. By adding an inverted U-shaped slot and an I-shaped slot on the back side, it can reduce the mutual coupling between the antenna elements. In addition, as shown in Antenna 3, introducing the step-shaped radiating elements can realize useful impedance bandwidth. The result indicates that the impedance bandwidth covers 2.33–2.68 GHz and 3.93–5.13 GHz, and the value of S_{21} adjacent to the defected ground structure decreases to below −24 dB and −17 dB in the 2.5 GHz and 4.5 GHz bands, respectively. Finally, a T-shaped slot and a G-shaped slot are incorporated at the bottom plane to reduce the mutual coupling to below −15 dB for the entire frequency band.

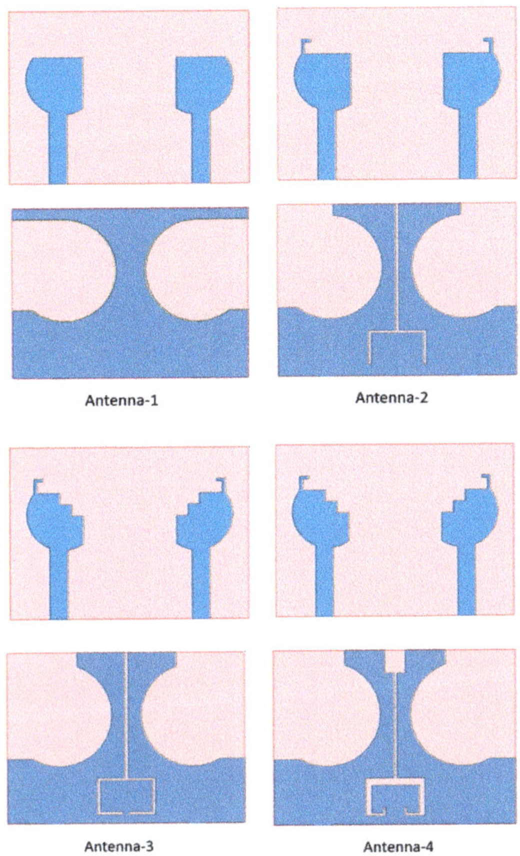

Figure 2. Design stages of the proposed design.

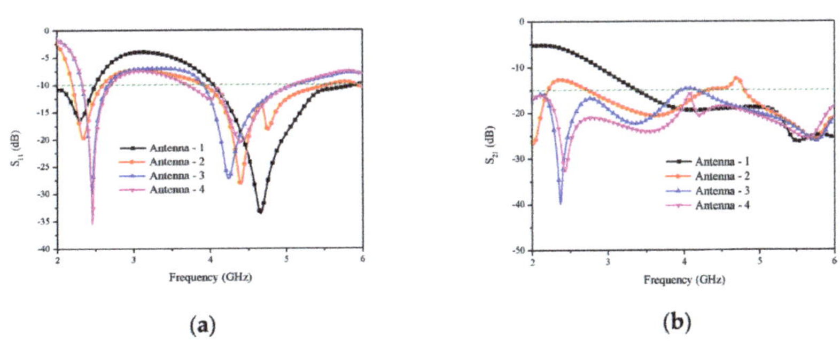

Figure 3. Design stage responses of the intended two-port MIMO antenna: (**a**) S_{11} responses and (**b**) S_{21} responses.

2.3. Parametric Study

The radiators can generate two resonances. The incomplete circular radiation patch generates the lower resonance of the antenna, and the L-shaped antenna branches can generate the higher resonance. Figure 4 describes the two resonance frequencies, which can be independently tuned by changing the values of R_2 and H_1. When the value of R_2 increases, the lower resonance shifts to the lower frequencies, as is shown in Figure 4a. When the value of H_1 increases, the ground current path increases, the lower resonance frequency remains at 2.5 GHz, the higher resonance frequency shifts to the lower frequencies, and the -10 dB bandwidth of the higher frequency band is reduced. The optimum values of R_2 and H_1 are 9 mm and 2.5 mm, respectively.

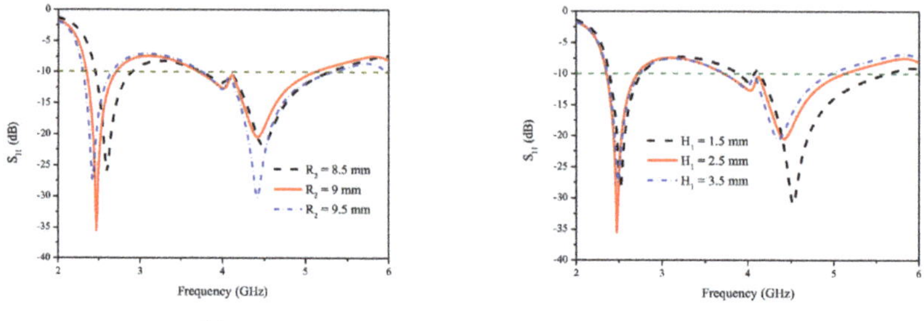

(a) **(b)**

Figure 4. Reflection coefficients of the proposed MIMO antenna (**a**) with different values of R_2 and (**b**) with different values of H_1.

2.4. Current Distribution

Figure 5 shows the simulated current distributions of the proposed antenna at 2.5 GHz and 4.7 GHz. After adding the DGS, one can see that the current from port 1 to port 2 is cut off. Therefore, high isolation is achieved because the surface current and near fields are concentrated within the decoupling structure.

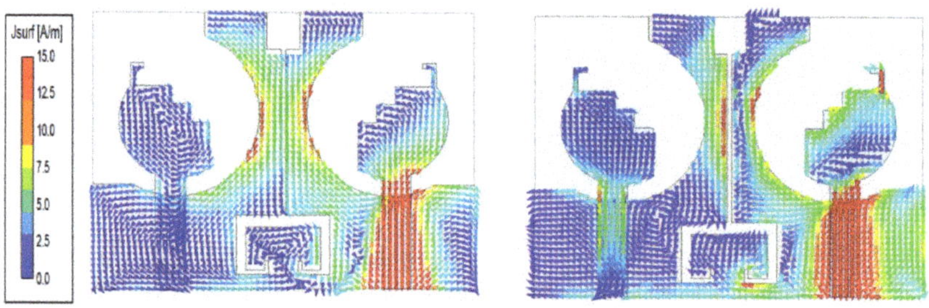

(a) **(b)**

Figure 5. Current distributions of the proposed MIMO antenna system at (**a**) 2.5 GHz and (**b**) 4.7 GHz.

When Ant_a is excited, the current is mainly concentrated on the edge of a circular patch at 2.5 GHz. While when it is working at 4.7 GHz, there is a strong current on the L-shaped branch and step-shaped part. Thus, each branch in the radiation elements is responsible for stimulating resonance. The formula controlling the relationship between excitation resonant frequency, antenna geometric parameters, and the physical characteristics is as follows:

$$f_r = \frac{c}{2L_e}\sqrt{\frac{2}{\epsilon_r + 1}} \quad (1)$$

where *Le* is the total electric length of the structural antenna elements.

3. Results and Discussion

3.1. S-Parameters

Figure 6 presents the simulated and measured S-parameters results for port 1, including S_{11} and S_{21}. It is exhibited that the measurement results are different from the simulation results because of SMA soldering and unavoidable tolerances in the fabrication and measurement process. To be specific, it is observed that the measured -10 dB impedance band can cover the n7 (2.5–2.69 GHz), n38 (2.57–2.62 GHz), n41 (2.496–2.690 GHz), and n79 bands (4.4–5 GHz) in Figure 6a. From the transmission curves, it can be seen that S_{21} is below −20 dB over the whole operating frequency band (2.0–6.0 GHz).

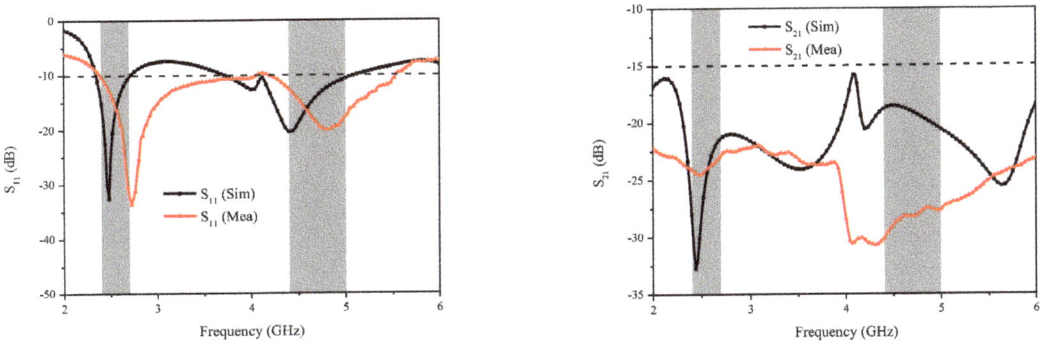

(a) **(b)**

Figure 6. Simulated and measured S-parameters of the proposed MIMO antenna system: (**a**) S_{11} and (**b**) S_{21}.

3.2. Radiation Pattern

To study the radiation mechanism of the antenna in-depth, Figure 7 presents the measured far-field radiation patterns in the H-plane and the E-plane at 2.5 GHz and 4.7 GHz, respectively. During the measurement and simulation, one port is excited, while the other one is matched with a 50 Ω load terminal. The proposed antenna has nearly stable omnidirectional radiation patterns in the H-plane at 2.5 GHz and 4.7 GHz. At 2.5 GHz, the E-plane is shown as a bidirectional model, while the radiation patterns appear deformed in the E-plane at 4.7 GHz. There is no effect on the radiation performance.

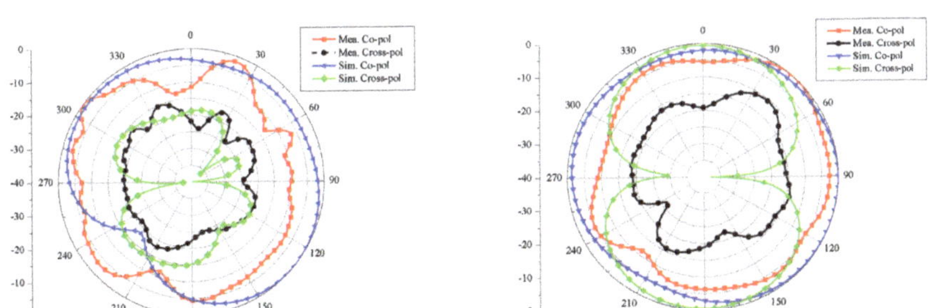

E-plane vs. H-plane

(**a**)

Figure 7. *Cont.*

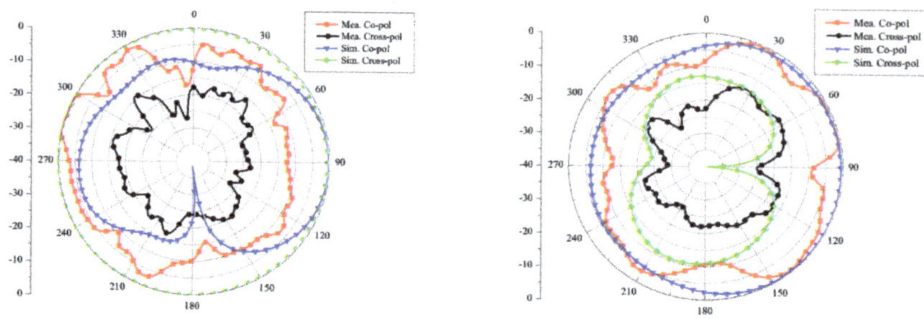

E-plane vs. H-plane

(**b**)

Figure 7. Two-dimensional radiation patterns at E-plane and H-plane for the intended two-port MIMO antenna at port-1: (**a**) 2.5 GHz and (**b**) 5.7 GHz.

4. The Performance Evaluation of MIMO Antenna Systems

4.1. The Realized Peak Gain and Efficiency

Gain, as one of the critical parameters of the antenna, is used to measure the antenna's ability to receive and transmit signals. Figure 8 shows the gain and radiation efficiency of the proposed MIMO antenna. It is presented that the peak gain of the proposed antenna is around 3 dBi for the 2.5 GHz band, and it is around 3.8 dBi for the 4.7 GHz band, while the measured radiation efficiency is higher than 62% and 66% in the operating bands, respectively.

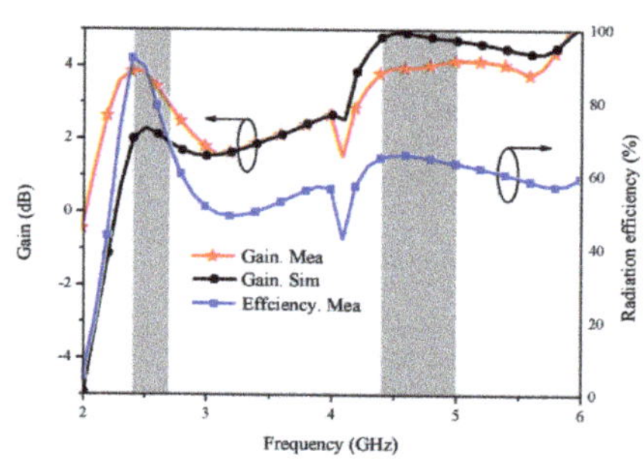

Figure 8. Gain and radiation efficiency of the proposed MIMO antenna.

4.2. Envelope Correlation Coefficient and Diversity Gain

In addition, gain, the envelope correlation coefficient, and diversity gain are also considered to be valuable parameters to evaluate the antenna system performance. ECC represents the degree of correlation between the antenna elements of a multiple antenna system. The calculation formulas of ECC calculated from the S-parameters or far-field radiation characteristics are shown in Equations (2) and (3) [16], respectively. However, only when the antenna efficiencies are nearly 100%, is Equation (2) accurate. Therefore, ECC is calculated using Equation (3) in this paper.

$$ECC = \frac{|S_{11}^{*}S_{12} + S_{22}^{*}S_{21}|^2}{(1 - |S_{11}|^2 + |S_{21}|^2)(1 - |S_{22}|^2 + |S_{12}|^2)} \tag{2}$$

$$\rho_e = \iint \frac{[\vec{F_1}(\theta,\varphi) \times \vec{F_2}(\theta,\varphi)]d\partial^2}{\iint \left|\vec{F_1}(\theta,\varphi)\right|^2 d\partial \iint \left|\vec{F_2}(\theta,\varphi)\right|^2 d\partial} \tag{3}$$

where '$F_i(\theta,\varphi)$' is the radiated field of the first antenna.

On the other hand, the DG acts as the other index to evaluate the MIMO antenna isolation performance. The diversity gain is calculated through the following Equation (4):

$$DG = 10\sqrt{1 - (ECC)^2} \tag{4}$$

Figure 9 depicts the measured values and simulated values of ECC and DG for the proposed MIMO antenna, respectively. It can be seen that the ECC is below 0.005, and the DG is greater than 9.65 dBi within the operating bands, which meets the engineering standard. A performance comparison of the proposed antenna with previous dual-band antennas is provided in Table 2. In this table, it can be observed that the presented dual-band MIMO antenna has achieved a more compact size, a lower ECC, and a higher isolation compared to those of the other referenced antennas.

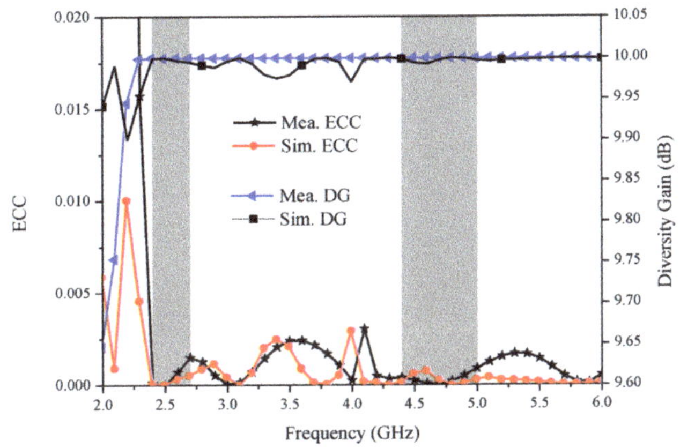

Figure 9. Measured and simulated ECC and DG of the proposed two-element MIMO antenna.

Table 2. Performance comparison with dual-band antennas.

Ref.	Operating Bands (GHz)	Isolation (dB)	ECC	Gain (dBi)	Size (mm^3)
[7]	2.25–2.9 5.05–6.025	>19.3	<0.03	2.4 and 3.8	50 × 50 × 1.6
[12]	2.23–2.46 3.22–4.04	>12	10^{-5} and 0.002	3.6 and 7.1	105 × 105 × 1.83
[13]	2.5–2.7 4.8–5.0	17.96 and 20.1	0.006 and 0.12	4.35 and 4.6	38.5 × 38.5 × 1.59
[15]	2.4–2.5 5.1–5.8	15	<0.2	1.5 and 1.5	38 × 38 × 1.6
[16]	2.3–2.5 5–5.2	20 and 20	<0.05	1.28 and 2.1	38 × 42 × 0.8
[17]	2.25–2.41 4.7–6.25	>18	<0.2	1.7 and 3	48 × 48 × 1.6
This work	2.34–2.71 3.72–5.10	21 and 18	<0.005 and <0.001	3 and 3.8	41 × 30 × 1.59

5. Conclusions

In this paper, a dual-band MIMO antenna with high isolation has been designed. The size of the proposed antenna is 41 × 30 × 1.59 $mm^{3,}$ and the impedance matching (which is better than −10 dB) covers 5G NR sub-6 GHz n7/n38/n41/n79. The high isolation of a value that is better than 20 dB is successfully realized by adding the defected ground structure and the slotted stub between the two antenna elements. Moreover, it is found that the antenna's lower gain and efficiency for the proposed system are 3 dBi and 62%, respectively. The measured ECC is less than 0.005, which meets the requirement of the MIMO systems. Therefore, the experimental results and the designed MIMO antenna structure exhibit that the presented antenna is preferred for 5G communication applications.

Author Contributions: Conceptualization, S.X.; methodology, J.C.; optimization, L.S.; investigation, Q.L.; measurement, L.S.; writing—original draft preparation, S.X.; writing—review and editing, Q.L. and G.L.; supervision and funding acquisition, G.L. All authors have read and agreed to the published version of the manuscript.

Funding: This work was funded in part by the Science and Technology Department of Zhejiang Province, under Grant No. LGG19F010009, the National Natural Science Foundation of China, under Grant No. 61671330, and Wenzhou Municipal Science and Technology Program, under Grant No. C20170005.

Institutional Review Board Statement: Not applicable.

Informed Consent Statement: Not applicable.

Data Availability Statement: Data are contained within the article.

Conflicts of Interest: The authors declare no conflict of interest.

References

1. Garg, P.; Jain, P. Isolation improvement of MIMO antenna using a novel flower shaped metamaterial absorber at 5.5 GHz WiMAX band. *IEEE Trans. Circuits Syst. II Express Briefs* **2020**, *67*, 675–679. [CrossRef]
2. Altaf, A.; Iqbal, A.; Smida, A.; Smida, J.; Althuwayb, A.A.; Hassan, K.S.; Alibakhshikenari, M.; Falcone, F.; Limiti, E. Isolation improvement in UWB-MIMO antenna system using slotted stub. *Electronics* **2020**, *9*, 1582. [CrossRef]
3. Zhou, J.Y.; Wang, Y.F.; Xu, J.M.; Du, C.Z. A CPW-fed UWB-MIMO antenna with high isolation and dual band-notched characteristic. *Prog. Electromagn. Res. M* **2021**, *102*, 27–37. [CrossRef]
4. Kulkarni, J.; Desaim, A.; Sim, C.Y.D. Wideband four-port MIMO antenna array with high isolation for future wireless systems. *AEU Int. J. Electron. Commun.* **2021**, *128*, 1434–8411. [CrossRef]
5. Sehrai, D.A.; Asifm, M.; Shoaib, N.; Ibrar, M.; Jan, S.; Alibakhshikenari, M.; Lalbakhsh, A.; Limiti, E. Compact quad-element high-isolation wideband MIMO antenna for mm-wave applications. *Electronics* **2021**, *10*, 1300. [CrossRef]
6. Liu, Y.; Yang, X.; Jia, Y. A low correlation and mutual coupling MIMO antenna. *IEEE Access* **2019**, *7*, 127384–127392. [CrossRef]
7. Yang, Q.C.; Zhang, C.B.; Cai, Q.B.; Loh, T.H.; Liu, G. A MIMO antenna with high gain and enhanced isolation for WLAN applications. *Appl. Sci.* **2022**, *12*, 2279. [CrossRef]
8. Pannu, P.; Devendra, S.K. Miniaturize four-port UWB-MIMO antenna with tri-notched band characteristics. *Microw. Opt. Technol. Lett.* **2021**, *63*, 1489–1498. [CrossRef]
9. Rekha, V.S.D.; Pardhasaradhi, P.; Madhav, B.T.P.; Devi, Y.U. Dual band notched orthogonal 4-element MIMO antenna with isolation for UWB applications. *IEEE Access* **2020**, *8*, 145871–145880. [CrossRef]
10. Huang, J.; Dong, G.; Cai, J.; Li, H.; Liu, G. A quad-port dual-band MIMO antenna array for 5G smartphone applications. *Electronics* **2021**, *10*, 542. [CrossRef]
11. Ren, Z.; Zhao, A. Dual-band MIMO antenna with compact self-decoupled antenna pairs for 5G mobile applications. *IEEE Access* **2019**, *7*, 82288–82296. [CrossRef]
12. De-sai, A.; Upadhyaya, T.; Palandoken, M.; Gocen, C. Dual band transparent antenna for wireless MIMO system applications. *Microw. Opt. Technol. Lett.* **2019**, *61*, 1845–1856. [CrossRef]
13. Kiani, S.H.; Altaf, A.; Abdullah, M.; Muhammad, F.; Blaauskas, T. Eight element side edged framed MIMO antenna array for future 5G smartphones. *Micromachines* **2020**, *11*, 956. [CrossRef] [PubMed]
14. Dong, G.T.; Huang, J.L.; Chen, Z.Z.; Liu, G. A compact planar dual band two-port MIMO antenna with high isolation and efficiency. *Int. J. RF Microw. Comput. Aided Eng.* **2022**, *32*, e23245. [CrossRef]
15. Thummaluru, S.R.; Kumar, R.; Chaudhary, R.K. Isolation and frequency reconfigurable compact MIMO antenna for wireless local area network aplications. *IET Microw. Antennas Propag.* **2019**, *13*, 519–525. [CrossRef]
16. Sipal, D.; Abegaonkar, M.P.; Koul, S.K. Highly isolated compact planar dual-band antenna with pattern diversity characteristic for MIMO terminals. *IEEE Antennas Wirel. Propag. Lett.* **2019**, *18*, 762–766. [CrossRef]
17. Naidu, P.V.; Nhanekula, M.B.; Almustafa, K.M. Design and performance analysis of MAZE shaped quad port ACS fed tri band MIMO antenna for V2V and multiband applications. *AEU Int. J. Electron. Commun.* **2021**, *134*, 153676. [CrossRef]
18. Kannappan, L.; Palaniswamy, S.K.; Wang, L.; Kanagasabai, M.; Kumar, S.; Alsath, M.G.N.; Rao, T.R. Quad-port multiservice diversity antenna for automotive applications. *Sensors* **2021**, *21*, 8238. [CrossRef] [PubMed]

19. Du, C.Z.; Zhao, Z.L. A CPW-fed dual-band MIMO antenna with enhanced isolation for 5G application. *Prog. Electromagn. Res. M.* **2020**, *98*, 11–20. [CrossRef]
20. Chen, H.D.; Tsai, Y.C.; Sim, C.Y.D.; Kuo, C. Broadband eight-antenna array design for Sub-6 GHz 5G NR bands metal-frame smartphone applications. *IEEE Antennas Wirel. Propag. Lett.* **2022**, *19*, 1078–1082. [CrossRef]

Disclaimer/Publisher's Note: The statements, opinions and data contained in all publications are solely those of the individual author(s) and contributor(s) and not of MDPI and/or the editor(s). MDPI and/or the editor(s) disclaim responsibility for any injury to people or property resulting from any ideas, methods, instructions or products referred to in the content.

micromachines

MDPI

Article

Low Profile Meandered Printed Monopole WiMAX/WLAN Antenna for Laptop Computer Applications

Killol Vishnuprasad Pandya

Electronics and Communication Engineering Department, Chandubhai S Patel Institute of Technology, CHARUSAT University, Gujarat 388421, India; killolpandya.ec@charusat.ac.in

Abstract: The research on wireless communication demands technology-based efficient radio frequency devices. A printed monopole dual-band antenna is designed and presented. The presented antenna exhibits a promising response with improved bandwidth and gain. The antenna radiates from 3.49 GHz to 3.82 GHz and from 4.83 GHz to 5.08 GHz frequencies with 3.7 dBi and 5.26 dBi gain, having a bandwidth of 9.09% and 5.06%, respectively. The novelty in the developed antenna is that resonating elements have been engineered adequately without the use of the additional reactive component. The cost-effective FR 4 laminate is utilized as a substrate. This structure exhibits an efficiency of over 83% for both resonances. The numerically computed results through simulations and measured results are found to be in good correlation. The aforesaid response from the antenna makes it an appropriate candidate for laptop computer applications.

Keywords: monopole antenna; multiband antenna; tablet computers; wideband antenna

Citation: Pandya, K.V. Low Profile Meandered Printed Monopole WiMAX/WLAN Antenna for Laptop Computer Applications. *Micromachines* **2022**, *13*, 2251. https://doi.org/10.3390/mi13122251

Academic Editor: Mark L. Adams

Received: 20 October 2022
Accepted: 14 December 2022
Published: 17 December 2022

Publisher's Note: MDPI stays neutral with regard to jurisdictional claims in published maps and institutional affiliations.

Copyright: © 2022 by the author. Licensee MDPI, Basel, Switzerland. This article is an open access article distributed under the terms and conditions of the Creative Commons Attribution (CC BY) license (https://creativecommons.org/licenses/by/4.0/).

1. Introduction

In the last few years, considerable growth has been observed in the utilization of mobile devices. Researchers have paid significant time to the development and exploration of conventional antennas. A low profile and stable monopole antenna for wireless wide area network communication is discussed in [1,2]. In [3], monopole multiband printed antenna is designed and analyzed for smart grid devices and laptops. The two monopole slots were created to have a balance between radiation efficiency and bandwidth. In embedded antennas, size miniaturization is very essential and needy. A systematic approach to fix a long monopole in limited space is by the implementation of branches. These branches of a particular shape could excite the desired resonant modes. The current density of the surface determines the radiation pattern and resonances. Notably, optimizing the dimensions of these strips shall make the structure appropriate for required applications. Lump elements could be incorporated with monopole strips to get resonance at targeted frequencies. However, lump elements may cause a decrement in antenna efficiency. Planar antennas have attractive design characteristics, such as ease of fabrication and a low profile. Due to these advantages, they are appropriate candidates for and it is the choice of researchers in handheld devices, such as laptops and tablet computers. [4–6]. A combination of metal loops and dielectric substrate exhibits a convenient alternative in monopole design [7–9]. The literature has shown that inverted F antenna with C-shaped radiator and meander shorting strips could exhibit dual resonance for wireless and wireless local area network (WLAN) applications [10]. Planar Inverted F Antenna (PIFA) also claims significant utilization in handheld devices [11]. Such a type of PIFA antenna requires considerable vertical space and they also have issues regarding mutual coupling with the substrate. A compact frequency reconfigurable antenna is developed and presented for computer tablet applications [12]. In this model, the RF switch is designed with shorting strips to alter the resonant modes. It is always a challenging issue to achieve optimum values of typical antenna parameters, such as bandwidth and gain together.

Electrically small antennas are preferable to be embedded in communication devices. However, these antennas always suffer from the issue of desired gain and efficiency at target resonance [13]. Metamaterial-inspired antennas could be the alternative to enhance gain with size miniaturization [14]. In this paper, it has been reported that a combination of Split Ring Resonator (SRR) and metamaterial offers multiband resonance achievement with adequate gain.

In the proposed antenna structure, FR4 material is utilized as a substrate, which is a dielectric material whose permittivity (εr) is 4.4. Though other materials having lesser permittivity exist in the market, FR4 is preferred because of ease of availability and cost-effectiveness. Due to the low cost, bulk production of this structure is feasible. The detailed antenna geometry is discussed in the Antenna Design section. The subsequent sections explain the comparison of simulated and measured results, parametric study analysis, fabricated antenna testing, and measurements. The final section concludes the conducted research.

Several techniques had been utilized to develop an antenna structure for mobile or tablet applications. The monopole antennas with meandered strips were discussed for laptop/tablet or mobile applications [14,15]. However, the size of these antennas is a challenging task to embed for WiMAX/WLAN applications. Many size reductions geometries have been reported for targeted applications. In [16], a uniplanar antenna was proposed to target UMTS/GSM and LTE operations. The researchers created a printed loop that formed a matching circuit for an antenna. Notably, a widened portion of a parasitic shorted strip, having a width of 3.4 mm, was provided to enhance the bandwidth for lower resonating frequencies. The presented structure utilized a combination of the shorted strip, a loop, and an inductive strip to get antenna response for target frequencies. Similar research has presented another RF structure for WWAN/LTE frequency applications [17]. As discussed previously, instead of developing widened strip, T shaped strip and U shaped strip were incorporated with a monopole antenna to meet the desired requirements. The combination of these strips significantly reduced the antenna size. In addition, the miniaturized structure could be easily mountable on tablet/laptop devices. A reconfigurable antenna, resonating for multiband frequencies was claimed in [18]. The RF switch was embedded with antenna geometry to alter the resonating frequencies of the lower band for four various working states. However, the designing concept is similar to having a couple of shorting strips as discussed earlier. The presented structure exhibits promising radiation efficiency and antenna gain. A couple of strips with a shorting strip were incorporated to excite the resonant modes. Additionally, the integration of metal components in the structure gave a fair rise in antenna efficiency and operating bandwidth. Furthermore, due to incorporation of resonating strips, the frequency of resonating band is shifted to the targeted frequencies without any additional reactive component. The additional reactive elements adds fabrication cumbersomeness. This is the novelty of a presented antenna.

2. Antenna Design and Geometry

Figure 1 illustrates the systematic design flow of the proposed antenna. The flow describes the steps that have been performed in a sequence to identify and rectify errors, if any.

The monopole antenna is demonstrated with metallic strips developed at the top of the structure as shown in Figure 2a that depicts the top view of an antenna. The close observation of the top view exhibits two stubs are provided at the end of strips to get resonance at desired frequencies. The microstrip line feed technique is used to energize the design. Figure 2b,c show the back view and trigonometric view of the model, respectively. By keeping the basic concepts of antenna radiation in mind, the geometry has been proposed. Many corners and branches in the conducting strips are provided to increase the current distribution. The response from an antenna due to the variation in dimensions of the ground plane and the addition of various metallic strips is analyzed in the parametric study section. In the isometric view, all layers are visible. The standard height of 1.6 mm is fixed for the substrate.

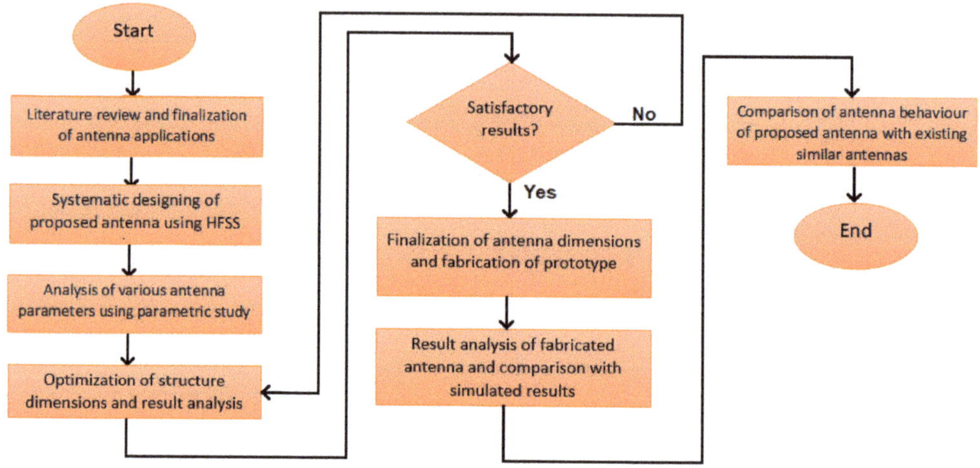

Figure 1. Development flow of proposed antenna.

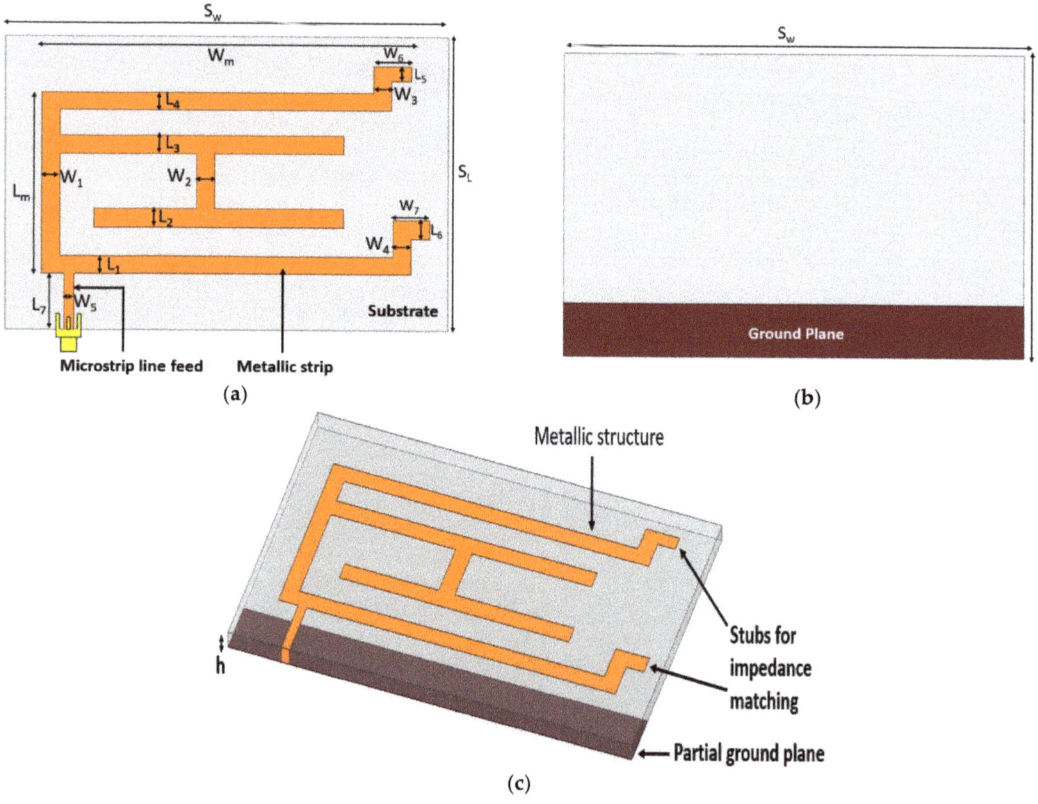

Figure 2. Proposed monopole antenna. (**a**) Top view (**b**) Back view (**c**) Trigonometric view.

Table 1 illustrates the dimensions of the proposed antenna. The dimensions are optimized to have acceptable impedance matching. By developing many stubs, the discontinuity over the entire structure was increased to receive optimal radiation.

Table 1. Antenna dimensions.

Notations	Dimensions (mm)	Notations	Dimensions (mm)
W_1	2.4	L_5	2
L_1	2.4	W_6	5
W_2	2.4	L_6	2.5
L_2	2.5	W_7	5
W_3	2.4	L_7	7.6
L_3	2.5	L_m	24.8
W_4	2.4	W_m	50
L_4	2.4	S_W	60
W_5	1.3	S_L	40

The fabricated antenna is presented in Figure 3. A SubMinature version A (SMA) connector is soldered with the microstrip line feed with possible accuracy to get satisfactory output. Figure 3a gives an idea of metallic structure at the top surface and Figure 3b shows the back view where the partial ground plane is visible. The proposed laptop antenna design is compact in comparison with another similar laptop/tablet antenna structure. The detailed comparison is presented in the last phase of the paper.

(a)

(b)

Figure 3. Fabricated monopole antenna. (**a**) Top view (**b**) Back view.

(a) Impedance matching for maximum power transmission

To elaborate the design of the presented antenna, Figure 4 shows the graph of input impedance over the targeted frequency spectrum. It is noticeable from the figure that for the frequencies between 3.49 GHz to 3.82 GHz and from 4.83 GHz to 5.08 GHz, the input impedance (Zin) is almost 50 Ω, which is expected. At similar frequency spans, the reactance is exactly 0 Ω. This proves excellent impedance matching for maximum power transfer.

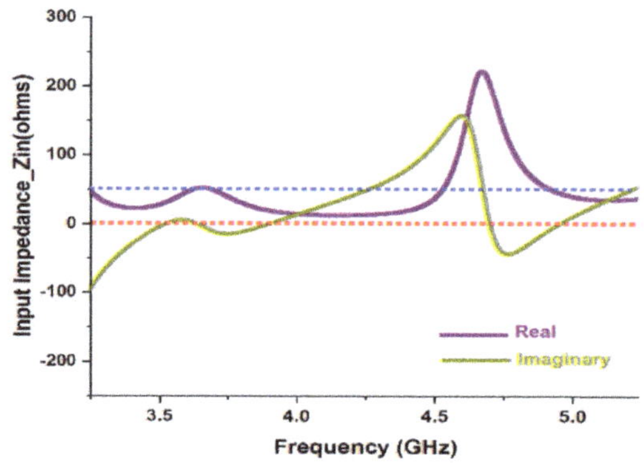

Figure 4. Graph of input impedance vs. frequency.

(b) ECD for the proposed antenna

Figure 5a gives the Equivalent Circuit Design (ECD) of the proposed antenna. The scattering matrix parameter S_{11} (dB) is shown in Figure 5b. This could be carried out by Agilent Advanced Design System (ADS) software. The circuit contains a parallel resistor-inductor-capacitor (RLC) network, which is connected with an input/output port at one end and a load resistor of 50 ohms at the other end.

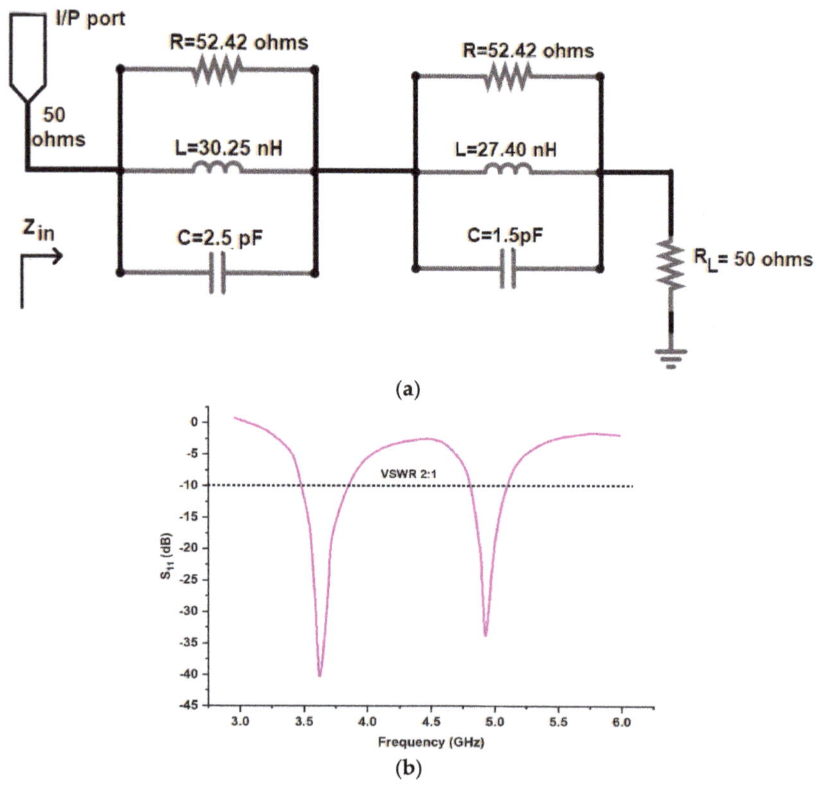

Figure 5. (**a**) Schematic ECD model and (**b**) S11 (dB) of the proposed antenna.

The values of RLC resonators could be obtained from the following equations [19,20].

$$Q_0 = \frac{f}{BW} \tag{1}$$

$$Q_0 = 2\pi RC \tag{2}$$

$$f = \frac{1}{2\pi\sqrt{LC}} \tag{3}$$

Here, f is resonating frequency, Q_0 is the quality factor of parallel RLC resonator, and R, L, and C are the resistor, inductor, and capacitor, respectively. The value of R could be found out using Figure 5 and is 52.42 Ω.

3. Parametric Study

To finalize the parameters of the proposed antenna, multiple iterations were performed. Figure 6 illustrates the comparison of antenna response for different antenna parameters values. The presented structure was analyzed for several dimensions of the ground plane. As the ground plane is made from copper, which is a conducting material that forms capacitor geometry with metallic structure at the top surface, it is very crucial to fix its dimensions. Figure 6a depicts the antenna model responses with a fully ground plane, half ground plane, and partial ground plane. The close observation shows, partial ground plane-based structure exhibits better output in terms of application requirements. The partial ground plane permits bidirectional radiation in the azimuth plane compared to the typical directional pattern. The metallic structure having open ends provide abundant opportunities for design tuning and response optimization. The strip line-based metallic structure is finalized after rigorous iterations.

From various iterations, the selected responses are shown in Figure 6b to explain the systematic development of antenna structure. As noticed, in the first iteration, the C-shaped and T-shaped metallic structure is developed and the response has been carried out. In addition to the first iteration, a T-shaped geometry is supported by a horizontal stripe in the second iteration. The proposed antenna structure has included extra stubs at the open ends of the c-shaped design to shift the return loss towards resonant frequencies. The base for two stubs was incorporated in the third iteration, which depicts the satisfactory result in terms of return loss. The fourth iteration exhibits output by the proposed antenna, which gives an optimum result at targeted resonances.

Figure 6c,d represents a comparison of reflection coefficient values for possible variations in microstrip feed width and length, respectively. Many iterations have been carried out where all exhibit acceptable values. However, the structure having a feed width of 1.3 mm and feed length of 7.6 mm shows adequate return loss. The horizontal stubs are provided at the open ends of the c-shaped geometry. The parametric study has been carried out on the width of the stub. By introducing a stub, the response could be achieved at the targeted frequency. Figure 6e illustrates the outputs for variation in the width of stubs. The verified dimensions of these widths are 3 mm, 5 mm, and 6 mm. The precise observation says, desired resonances could be obtained using a 5 mm width dimension. Figure 6f gives an analysis of reflection coefficient values for various dimensions of L2 and L3. The graph depicts, satisfactory output has been maintained using a length dimension of 2.4 mm.

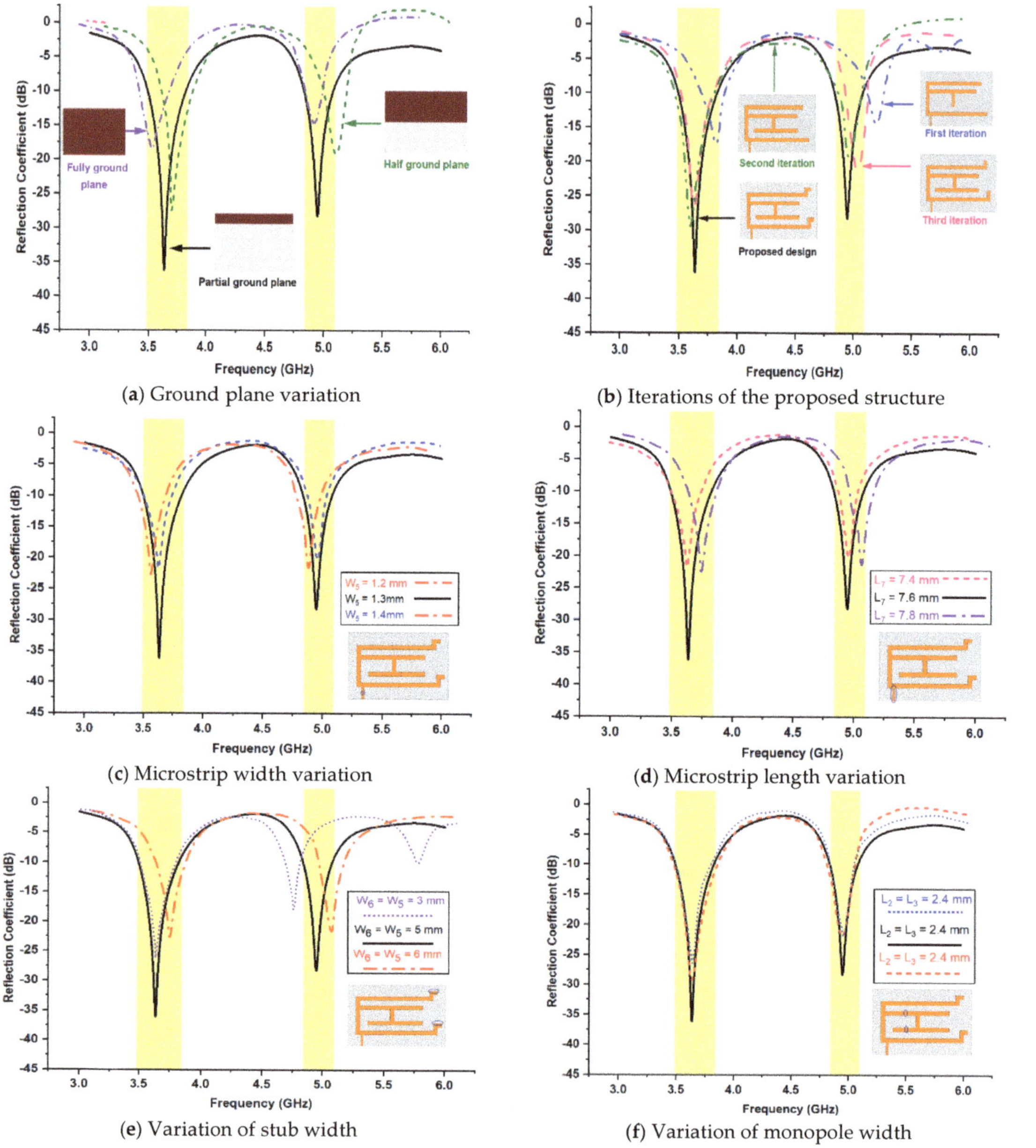

(**a**) Ground plane variation

(**b**) Iterations of the proposed structure

(**c**) Microstrip width variation

(**d**) Microstrip length variation

(**e**) Variation of stub width

(**f**) Variation of monopole width

Figure 6. Parametric study of return loss from possible iteration. (**a**) Ground plane variation (**b**) Iterations of the proposed structure (**c**) Microstrip width variation (**d**) Microstrip length variation (**e**) Variation of stub width (**f**) Variation of monopole width.

4. Simulated and Measured Results

The software-generated results for return loss are carried out using FEM-based High-Frequency Structure Simulator (HFSS) software (Ansys, Bangalore, India). The fabricated antenna has been tested through VNA 9912A (Keysight, Bangalore, India). The comparison of simulated, measured and Equivalent Circuit Design (ECD) results of reflection coefficient values for certain frequency variation is illustrated in Figure 7. For the majority of frequencies, the overlapping of lines between simulated and measured results could be observed. The measured result follows the software-generated result, which is the indication of enhanced impedance matching and high accuracy of developed antenna. The antenna has two modes of resonance having the center frequency of 3.63 GHz and 4.94 GHz. These frequencies fall under the S band and C band frequency range. The measured bandwidth for targeted frequency bands is 9.09% (3.49 GHz–3.82 GHz) and 5.06% (4.83 GHz–5.08 GHz).

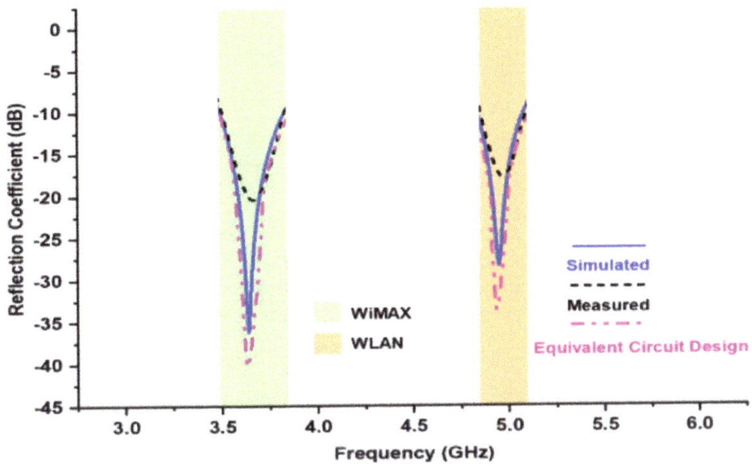

Figure 7. Comparison between simulated, measured, ECD reflection coefficient values.

Figure 8 depicts the current distribution on the monopole of the proposed antenna. The various color indicated current density at a particular portion of the monopole. Impedance matching plays a vital role in the current distribution. Additionally, the antenna radiation is proportional to the current distribution on the surface. In this figure, the majority of current distribution could be visible over the outer C shape and the shorting strips. The microstrip feed is also having the adequate current flow, which is a prime requirement. Due to the fringing filed pattern some of the portion still has a blue color, which indicates less current distribution, however collective structure has satisfactory current distribution. The fabricated antenna was tested and analyzed by an anechoic chamber. Figure 9a,b shows an antenna inside the anechoic chamber for E-field and H-field radiation pattern measurement, respectively. The size of the anechoic chamber is 5 m × 5 m × 5 m.

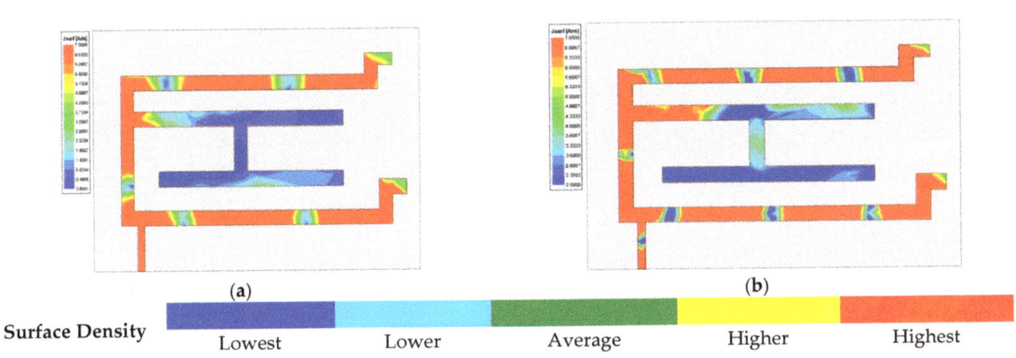

Figure 8. Current density at (**a**) 3.63 GHz and (**b**) 4.94 GHz frequencies.

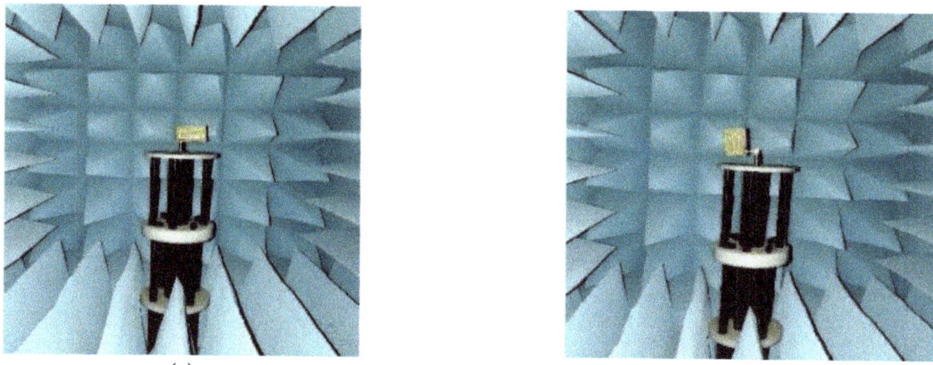

(a) (b)

Figure 9. Testing set-up of the fabricated antenna in an anechoic chamber. (**a**) E-field (**b**) H-field.

The two dimensions normalized radiation characteristics of the presented monopole antenna for azimuth (E-field) and elevation (H-field) plane are obtained and illustrated by Figure 10. It exhibits gain values of 3.7 dBi and 5.26 dBi, respectively. The simulated radiation pattern and measured radiation patterns are in close correlation.

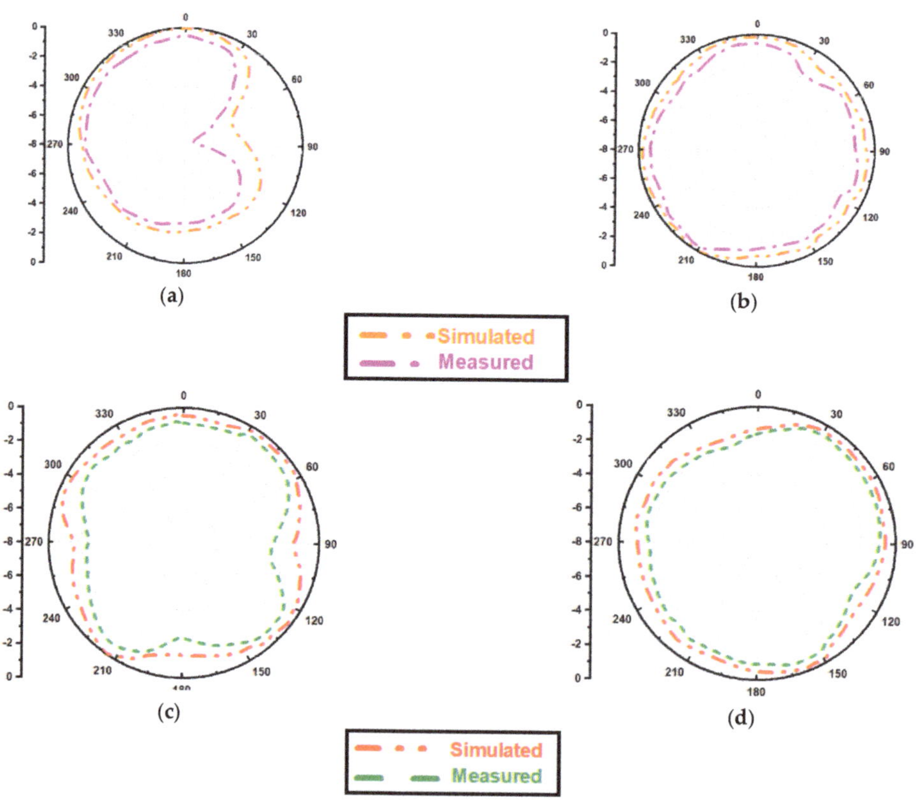

Figure 10. Normalized E-field [(**a**,**b**)] and H-field [(**c**,**d**)] radiation pattern at 3.63 GHz and 4.94 GHz frequencies.

Figure 11 depicts co-polarization and cross-polarization of the far-field pattern. The structure exhibits a satisfactory gain pattern having co and cross-polarization isolation is more than 17 dBi. The three dimensions polar plot for desired frequencies are shown with the help of Figure 12. The given radiation pattern primarily exhibits a couple of important parameters. The peak gain values are 3.7 dBi and 5.26 dBi at 3.63 GHz and 4.94 GHz frequencies, respectively. In addition, the far-field radiation is nearly omnidirectional.

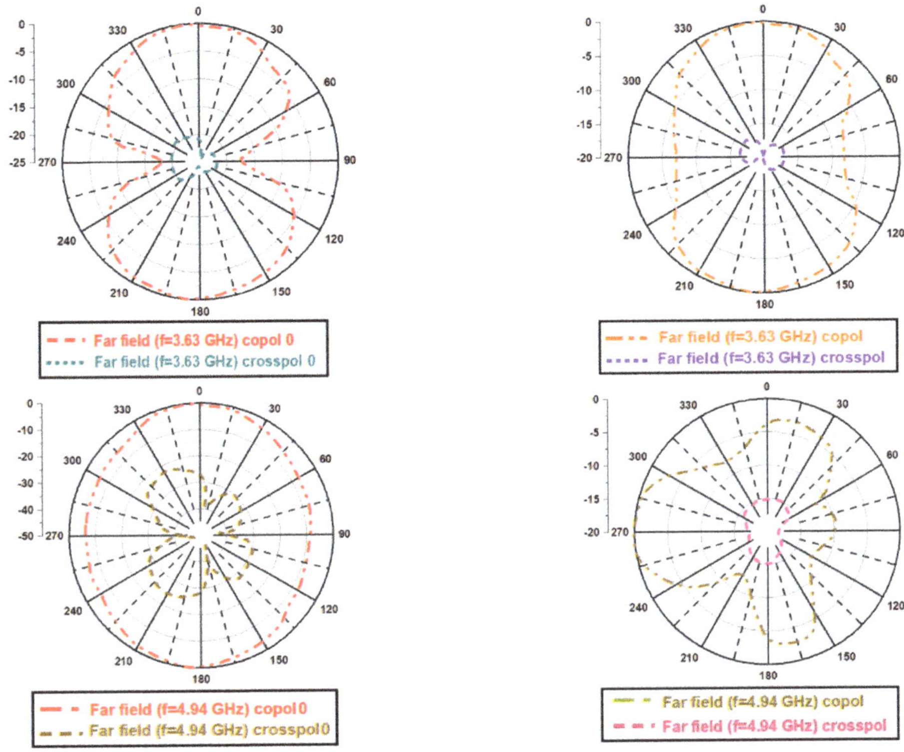

Figure 11. Co polarization and cross-polarization.

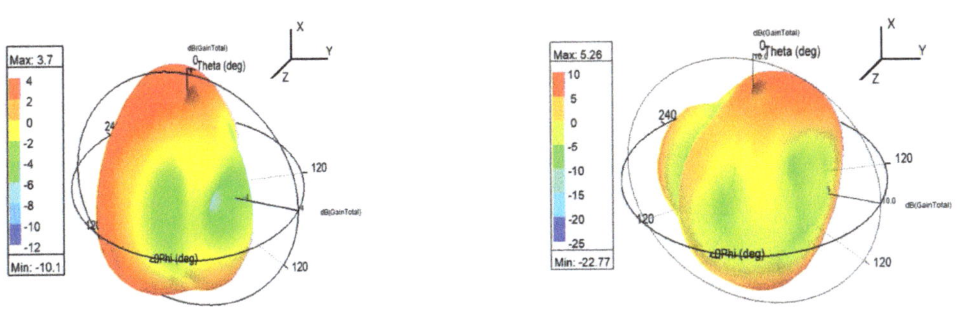

(a) (b)

Figure 12. 3-Dimension radiation pattern. (**a**) 3.63 GHz frequency (**b**) 4.94 GHz frequency.

As illustrated in Figure 13, as the frequency increases the directivity of the presented monopole antenna increases. As a result of the increment in directivity, the gain value also increases. Due to a significant reduction in conduction loss, the gain could be further optimized [18]. The measured gain is also shown in a similar graph to observe the correlation between simulated and measured gain values. The measured value varies in comparison with the simulated due to instrumentation error. The gain could be obtained from the following formula.

$$G_2(\text{dB}) = 20\log_{10}\left(\frac{4\pi r}{\lambda}\right) + 10\log_{10}\left(\frac{P_2}{P_1}\right) - G_1(\text{dB}) \quad (4)$$

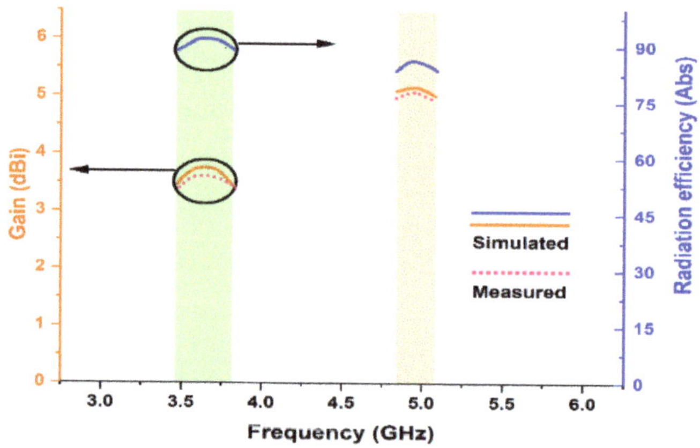

Figure 13. Graph of gain and radiation efficiency vs. frequency.

Here, λ is wavelength, $G_2(\text{dB})$ is the gain of the proposed structure, and $G_1(\text{dB})$ is the gain of the reference antenna. Similarly, r is the distance between the proposed antenna and the reference antenna. P_1 and P_2 are transmitted and receive power of reference and proposed antenna.

The efficiency is also presented in Figure 10. The radiation efficiency for 3.63 GHz and 4.94 GHz frequencies are 91.47% and 85.61%, respectively.

Figure 14 depicts the anechoic chamber setup. The antenna is placed on the receiver side where the reference horn antenna is fixed at the transmitter end. The anechoic chamber has absorbers at the inner surface to provide the ideal atmosphere for antenna measurement.

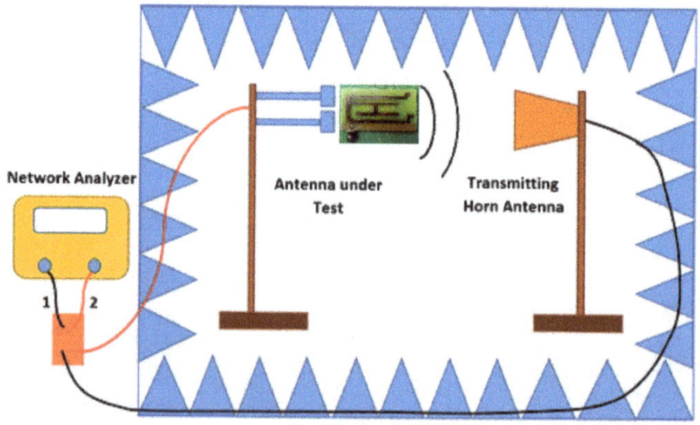

Figure 14. Anechoic chamber set up.

The proposed work simulated results are reported in Table 2. All dimensions are shown in terms of λ. Here also, λ is wave length. The proposed structure is compact with respect to other similar structures and also exhibits moderate gain and sufficient efficiency.

Table 2. Comparison of proposed monopole antenna with other similar structures.

Citation	Resonating Frequencies	Antenna Dimensions	Simulated Gain (dBi)	Simulated Efficiency (%)	Simulated S_{11} (dB)	Simulated FBW (%)
[1]	(698–960) MHz and (1710–2690) MHz	60 × 50 × 1.6 (0.48λ × 0.4λ × 0.01λ)	1.82 and 2.02	50	−28 and −16	3.7 and 12, 8.9
[3]	2.4, 3.7 (GHz)	50 × 200 × 1.6 (0.4λ × 1.61λ × 0.012λ)	1.89 and 0.97	Not given	−22.65	5.42 and 5.40
[15]	0.85, 0.92, 1.79 (GHz)	60 × 200 × 0.8 (0.17λ × 0.56λ × 0.002λ)	0.6, 0.5 and 1.2	60–72	−28.10, −24.65, −16.65	10, 4.6 and 8.5
[16]	0.83, 1.95, 2.35, 2.66 (GHz)	60 × 200 × 4 (0.16λ × 5.55λ × 0.01λ)	−0.9, 0.1, −0.95 and −0.98	68	−25.01, −21, −15, −29.66	1.8, 2.6, 3.04 and 2.0
[21]	(0.698–1.10) GHz and (1.64–2.83) GHz	60 × 118 × 0.8 0.02λ × 0.29λ × 0.002λ)	1.53, 2.91 and 0.85	Not given	−23, −14	2.75 and 5.56
[22]	(800–1300) MHz, (1710–2325) MHz	60 × 115 × 1.6 (0.22λ × 0.42λ × 0.005λ)	Not given	65.21, 70.26	−19.67, −22.58	6.25 and 20.45
[23]	(660–1065) MHz, (1665–3000) MHz	60 × 105 × 1.6 (0.19λ × 0.33λ × 0.005λ)	0.5 and 2	60, 84	−21.75 and −19.15	22.15 and 32.55
[24]	(750–1040) MHz, (1635–2485) MHz	60 × 115 × 1.6 (0.17λ × 0.32λ × 0.004λ)	3.25 and 3.09	80,82	−30.77 and −22	10 and 9.18
Proposed structure	(3.49–3.82) GHz and (4.83–5.08) GHz	60 × 40 × 0.8 (0.72λ × 0.48λ × 0.04λ)	3.7 and 5.26	91.47, 85.61	−38, −29.05	3.3 and 7

It has been observed that electrically miniaturized antenna could be demonstrated by depletion of oxide layers over the antenna surface. However, the gain could be compromised in such cases [25].

5. Conclusions

A printed monopole antenna of size 60 mm × 40 mm is designed, developed, fabricated, and tested. The claimed structure gives response at 3.63 GHz and 4.64 GHz frequencies, which are under the S band and C band frequency range. The simulated results and measured results have a significant correlation. The presented design offers a bandwidth of 9.09 % and 5.06 % with a moderate gain of 3.7 dBi and 5.26 dBi, respectively, for desired frequency bands. Here, the conducting strips are added to excite the resonating modes without any additional reactive component. The reactive component adds cumbersomeness in the manufacturing process, whereas a stub-based design avoids fabrication issues. The discussed antenna also illustrates an adequate response for other key parameters, such as return loss and radiation pattern. The proposed structure is well suitable for WiMAX and WLAN applications.

Funding: The research was funded by Charotar University of Science and Technology (CHARUSAT) under the CHARUSAT Seed Grant for research (Project No 14). The research was performed and carried out in the Electronics and Communication Engineering department, Chandubhai S. Patel Institute of Technology, CHARUSAT University, Gujarat, India.

Data Availability Statement: Not applicable.

Conflicts of Interest: The author declares no conflict of interest.

References

1. Chen, S.-C.; Huang, C.-C.; Cai, W.-S. Integration of a low-profile, long-term evolution/wireless wide area network monopole antenna into the metal frame of tablet computers. *IEEE Trans. Antennas Propag.* **2017**, *65*, 3726–3731. [CrossRef]
2. Pandya, A.; Upadhyaya, T.K.; Pandya, K. Tri-Band Defected Ground Plane Based Planar Monopole Antenna for Wi-Fi/WiMAX/WLAN Applications. *Prog. Electromagn. Res. C* **2021**, *108*, 127–136. [CrossRef]
3. Patel, H.; Upadhyaya, T.K. Printed Multiband Monopole Antenna for Smart Energy Meter/WLAN/WiMAX Applications. *Prog. Electromagn. Res.* **2020**, *89*, 43–51. [CrossRef]

4. Lu, J.-H.; Wang, Y.-S. Planar small-size eight-band LTE/WWAN monopole antenna for tablet computers. *IEEE Trans. Antennas Propag.* **2014**, *62*, 4372–4377. [CrossRef]
5. Ban, Y.-L.; Sun, S.-C.; Li, J.L.-W.; Hu, W. Compact coupled-fed wideband antenna for internal eight-band LTE/WWAN tablet computer applications. *J. Electromagn. Waves Appl.* **2012**, *26*, 2222–2233. [CrossRef]
6. Sim, C.-Y.; Chen, C.-C.; Li, C.-Y.; Ge, L. A novel uniplanar antenna with dual wideband characteristics for tablet/laptop applications. *Int. J. RF Microw. Comput.-Aided Eng.* **2017**, *27*, e21145. [CrossRef]
7. Wong, K.-L.; Lee, L.-C. Multiband printed monopole slot antenna for WWAN operation in the laptop computer. *IEEE Trans. Antennas Propag.* **2009**, *57*, 324–330. [CrossRef]
8. Chiu, C.-W.; Chang, C.-H.; Chi, Y.-J. Multiband folded loop antenna for smartphones. *Prog. Electromagn. Res.* **2010**, *102*, 213–226. [CrossRef]
9. Subbaraj, S.; Kanagasabai, M.; Alsath, M.G.N.; Ganesan, G.; Selvam, Y.P.; Kingsly, S. Compact multiservice monopole antenna for tablet devices. *Int. J. Electron.* **2018**, *105*, 1374–1387. [CrossRef]
10. Liu, H.-W.; Lin, S.-Y.; Yang, C.-F. Compact inverted-F antenna with meander shorting strip for laptop computer WLAN applications. *IEEE Antennas Wirel. Propag. Lett.* **2011**, *10*, 540–543.
11. Naser-Moghadasi, M.; Mansouri, Z.; Sharma, S.; Zarrabi, F.B.; Virdee, B.S. Low SAR PIFA antenna for wideband applications. *IETE J. Res.* **2016**, *62*, 564–570. [CrossRef]
12. Ban, Y.-L.; Sun, S.-C.; Li, P.-P.; Li, J.L.-W.; Kang, K. Compact eight-band frequency reconfigurable antenna for LTE/WWAN tablet computer applications. *IEEE Trans. Antennas Propag.* **2013**, *62*, 471–475. [CrossRef]
13. Ntaikos, D.K.; Bourgis, N.K.; Yioultsis, T.V. Metamaterial-based electrically small multi-band planar monopole antennas. *IEEE Antennas Wirel. Propag. Lett.* **2011**, *10*, 963–966. [CrossRef]
14. Pandya, A.; Upadhyaya, T.K.; Pandya, K. Design of Metamaterial Based Multilayer Antenna for Navigation/WiFi/Satellite Applications. *Prog. Electromagn. Res. M* **2021**, *99*, 103–113. [CrossRef]
15. Upadhyaya, T.K.; Desai, A.; Patel, R. Design of printed monopole antenna for wireless energy meter and smart applications. *Prog. Electromagn. Res. Lett.* **2018**, *77*, 27–33. [CrossRef]
16. Lu, J.-H.; Wang, Y.-S. Internal uniplanar antenna for LTE/GSM/UMTS operation in a tablet computer. *IEEE Trans. Antennas Propag.* **2013**, *61*, 2841–2846. [CrossRef]
17. Lu, J.-H.; Tsai, F.-C. Planar internal LTE/WWAN monopole antenna for tablet computer application. *IEEE Trans. Antennas Propag.* **2013**, *61*, 4358–4363. [CrossRef]
18. Vahora, A.; Pandya, K. Implementation of the cylindrical dielectric resonator antenna array for Wi-Fi/wireless LAN/satellite applications. *Prog. Electromagn. Res. M* **2020**, *90*, 157–166. [CrossRef]
19. Kulkarni, J. Multi-band printed monopole antenna conforming bandwidth requirement of GSM/WLAN/WiMAX standards. *Electromagn. Res. Lett.* **2020**, *91*, 59–66. [CrossRef]
20. Kulkarni, J.; Kulkarni, N.; Desai, A. Development of "H-Shaped" monopole antenna for IEEE 802.11 a and HIPERLAN 2 applications in the laptop computer. *Int. J. RF Microw. Comput. -Aided Eng.* **2020**, *30*, e2223. [CrossRef]
21. Zong, W.H.; Yang, X.M.; Xiao, X.; Li, S.D.; Wei, X.Y.; Jin, Z.J.; Qu, X.Y. A wideband antenna with circular and rectangular shaped slots for mobile phone applications. *Int. J. Antennas Propag.* **2016**, 2975425. [CrossRef]
22. Chang, C.-H.; Wei, W.-C.; Ma, P.-J.; Huang, S.-Y. Simple printed WWAN monopole slot antenna with parasitic shorted strips for the slim mobile phone application. *Microw. Opt. Technol. Lett.* **2013**, *55*, 2835–2841. [CrossRef]
23. Deng, C.; Li, Y.; Zhang, Z.; Feng, Z. Planar printed multi-resonant antenna for octa-band WWAN/LTE mobile handset. *IEEE Antennas Wirel. Propag. Lett.* **2015**, *14*, 1734–1737. [CrossRef]
24. Wong, K.-L.; Lin, P.-W.; Chang, C.-H. Simple printed monopole slot antenna for penta-band wireless wide area network operation in the mobile handset. *Microw. Opt. Technol. Lett.* **2011**, *53*, 1399–1404. [CrossRef]
25. Upadhyaya, T.; Desai, A.; Patel, R.; Patel, U.; Kaur, K.P.; Pandya, K. Compact transparent conductive oxide based dual band antenna for wireless applications. In Proceedings of the 2017 Progress in Electromagnetics Research Symposium-Fall (PIERS-FALL), Singapore, 19–22 November 2017; pp. 41–45.

micromachines

Article

A Leaky-Wave Analysis of Resonant Bessel-Beam Launchers: Design Criteria, Practical Examples, and Potential Applications at Microwave and Millimeter-Wave Frequencies

Edoardo Negri [1], Walter Fuscaldo [2,*], Paolo Burghignoli [1] and Alessandro Galli [1]

1 Department of Information Engineering, Electronics and Telecommunications, Sapienza University of Rome, 00184 Rome, Italy
2 Istituto per la Microelettronica e Microsistemi, Consiglio Nazionale delle Ricerche, 00133 Rome, Italy
* Correspondence: walter.fuscaldo@cnr.it

Abstract: Resonant Bessel-beam launchers are low-cost, planar, miniaturized devices capable of focusing electromagnetic radiation in a very efficient way in various frequency ranges, with recent increasing interest for microwave and millimeter-wave applications (i.e., 3–300 GHz). In recent years, various kinds of launchers have appeared, with different feeding mechanisms (e.g., coaxial probes, resonant slots, or loop antennas), field polarization (radial, azimuthal, and longitudinal), and manufacturing technology (axicon lenses, radial waveguides, or diffraction gratings). In this paper, we review the various features of these launchers both from a general electromagnetic background and a more specific *leaky-wave* interpretation. The latter allows for deriving a useful set of *design rules* that we here show to be applicable to any type of launcher, regardless its specific realization. Practical examples are discussed, showing a typical application of the proposed design workflow, along with a possible use of the launchers in a modern context, such as that of wireless power transfer at 90 GHz.

Keywords: Bessel beams; leaky waves; wireless power transfer; metasurfaces

Citation: Negri, E.; Fuscaldo, W.; Burghignoli, P.; Galli, A. A Leaky-Wave Analysis of Resonant Bessel-Beam Launchers: Design Criteria, Practical Examples, and Potential Applications at Microwave and Millimeter-Wave Frequencies. *Micromachines* **2022**, *13*, 2230. https://doi.org/10.3390/mi13122230

Academic Editors: Trushit Upadhyay and Hari Shankar Singh

Received: 14 November 2022
Accepted: 10 December 2022
Published: 15 December 2022

Publisher's Note: MDPI stays neutral with regard to jurisdictional claims in published maps and institutional affiliations.

Copyright: © 2022 by the authors. Licensee MDPI, Basel, Switzerland. This article is an open access article distributed under the terms and conditions of the Creative Commons Attribution (CC BY) license (https://creativecommons.org/licenses/by/4.0/).

1. Introduction

Any kind of wave (light, sound, etc.), initially confined to a finite area on a transverse plane, will be subject to diffractive spreading as it propagates outward from the source to free space. As a result, diffraction effects limit all applications in which it is required to maintain a spatial transverse confinement of a *beam* (i.e., a monochromatic electromagnetic wave with a well defined propagation axis) over a considerable distance. Such applications comprise but are not limited to wireless power transfer, free-space communications, medical imaging, radiometry, etc. For this reason, many efforts were made to find wave solutions that do not undergo diffraction.

Localized waves or *nondiffracting waves* are the largest family of *diffraction-free* solutions to the wave equation. As the name suggests, these waves ideally do not diffract, thus they maintain their transverse profile for an infinite distance from the radiating aperture. In principle, this remains true only for an infinite-size radiating aperture. Conversely, for a truncated radiating aperture, the localized waves can resist diffraction only up to a finite distance [1,2].

Bessel beams are one of the most known examples of localized waves. Bessel beams are monochromatic solutions of the Helmholtz equation in a cylindrical reference frame with very interesting focusing and limited-diffraction properties. These features attracted much interest for different applications in optics during the 1980s [3]. Currently, Bessel beams are studied and analyzed in the microwave and millimeter-wave ranges due to their usage in near-field focusing applications [4]. In these frequency ranges, Bessel beams are typically generated through planar structures, also known as *Bessel-beam launchers*. Such launchers are further distinguished between the *resonant* kind and the *wideband* kind. The former are realized through compact planar devices with only a few wavelengths of lateral size; they typically feature a narrow-band behavior due to their resonant character [5,6]. The latter, instead, feature a wider bandwidth at the expense of a lateral size that can be as large as several wavelengths [7–9]. Both kinds have been experimentally validated [6,7,10], but here we will deal only with the resonant type. Depending on the application, such as in WPT links for wearable devices, resonant Bessel-beam launchers are preferred to their wideband counterpart thanks to their compact implementation. Due to the lack of a comprehensive analysis of such devices in the available literature, only resonant Bessel-beam launchers are analyzed in this work (the interested reader can find information about wideband Bessel-beam launchers in, e.g., [4]).

Resonant Bessel-beam launchers can be realized in different ways, although they all exhibit azimuthal symmetry and share the same architecture: a circular metallic cavity excited by a simple dipole-like source at the center, whose upper plate is replaced by a partially reflecting sheet (PRS). The cavity height h plays a key role. When the cavity height is deeply subwavelength, $h \ll \lambda$, Bessel-beam launchers require a capacitive PRS and can only work with the *fundamental*, zeroth-order transverse-magnetic (TM) (with respect to the vertical z-axis) leaky mode [11]. When the cavity height is half a wavelength, $h \simeq \lambda/2$, the PRS can be either inductive or capacitive [11], and the launchers can work with either higher-order transverse-electric (TE) or TM leaky modes. Because ohmic losses in waveguide-like devices as Bessel-beam launchers decrease with the cavity height [12], higher-order resonant Bessel-beam launchers are usually preferred at millimeter-wave or sub-millimeter wave frequencies, where dissipative effects in metals and dielectrics may become important. In this paper, we will thus focus on such kinds of launchers.

Higher-order resonant Bessel-beam launchers can further be divided in two different kinds according to their feeding mechanism. In particular, if a vertical magnetic or electric dipole as source is used, then a TE- [13,14] or TM-polarized [6] Bessel beam is generated.

In the available literature, resonant Bessel-beam launchers in the microwave and millimeter-wave frequency range (i.e., 3–300 GHz) have been widely analyzed as standalone devices [5,6,13,15] and as elements in a WPT link [16,17], but a comprehensive analysis of this kind of launchers in terms of polarization, design, and physical implementations is still lacking. The main purpose of this work is indeed to provide a simple, yet rigorous step-by-step design workflow for differently polarized resonant Bessel-beam launchers based on leaky-wave theory. Theoretical results are exploited to derive design guidelines that are here applied for the physical implementation of an original device operating at 90 GHz. The radiating performance of the device is then corroborated through a comparison between full-wave and numerical results based on leaky-wave theory. Finally, different applications for Bessel beams are presented. In particular, the performance of a WPT link at 90 GHz realized through the resonant Bessel-beam launchers designed in this work is shown.

The paper is organized as follows. In Section 2, the theoretical framework underlying the physics and working principle of Bessel beams is first described. In Section 3, focused radiation from resonant Bessel-beam launchers is explained under a leaky-wave interpretation. This powerful analytical model allows for deriving simple design criteria to realize launchers with certain desired focusing properties. In Section 4, it is shown how the general framework developed in Sections 2 and 3 can fruitfully be applied to design practical devices that can easily be microfabricated through printed-circuit board (PCB) and/or standard photolithographic techniques, depending on the targeted frequency range. In Section 5, a brief discussion is provided about the possible application scenarios of the proposed devices. Finally, conclusions are drawn in Section 6.

2. Theoretical Framework

We follow [1] and start from the homogeneous wave equation in a cylindrical reference frame (ρ, ϕ, z) under the hypothesis of axial symmetry $(\partial/\partial\phi = 0)$:

$$\left(\frac{\partial^2}{\partial\rho^2} + \frac{1}{\rho}\frac{\partial}{\partial\rho} + \frac{\partial^2}{\partial z^2} - \frac{1}{c_0^2}\frac{\partial^2}{\partial t^2}\right)\psi(\rho, z, t) = 0 \tag{1}$$

where c_0 is the speed of light in vacuum and ψ is a component of the electromagnetic field or potential. In order to find a general solution to (1), a spectral method is applied: a Hankel transform with respect to ρ, and two Fourier transforms with respect to z and t lead to the following integral expression (a time dependence $e^{j\omega t}$ of the potential is assumed and suppressed throughout the paper):

$$\psi(\rho, z, t) = \int_{-\infty}^{+\infty}\int_{-\infty}^{+\infty}\int_{-\infty}^{+\infty} k_\rho J_0(k_\rho\rho)e^{-jk_z z}e^{j\omega t}\tilde{\psi}(k_\rho, k_z, \omega)dk_\rho dk_z d\omega \tag{2}$$

where k_ρ and k_z are the radial and longitudinal wavenumbers, respectively, ω is the angular frequency, and $\tilde{\psi}(k_\rho, k_z, \omega)$ is the generalized transform of $\psi(\rho, z, t)$. One order of integration can be skipped by exploiting the separation relation $k_0^2 = k_\rho^2 + k_z^2$, where $k_0 = \omega/c_0$ is the vacuum wavenumber. The integral in (3) can then be expressed in terms of an arbitrary spectral function $S(k_\rho, \omega)$ as:

$$\psi(\rho, z, t) = \int_0^{+\infty}\int_{-k_0}^{+k_0} k_\rho J_0(k_\rho\rho)e^{-jk_z z}e^{j\omega t}S(k_\rho, \omega)dk_\rho d\omega \tag{3}$$

where we have limited the domain of ω and k_ρ to consider only positive frequencies $\omega \in \mathbb{R}^+$ and fast, radiating waves $k_\rho \leq k_0$. The wave solution in (3) is rather general and expresses any solution of the Helmholtz equation in an axially symmetric reference frame, as a superposition of radiating plane waves. In order to obtain a localized solution, a linear coupling between ω and k_ρ has to be imposed [18]. This constraint is immediately obtained by enforcing the propagation-invariant property (which distinguishes a *localized* solution from any wave solution) to the wave solution $\psi(\rho, z, t) = \psi(\rho, t - z/v_z)$, where v_z is a constant (which implies rigid transport) representing the wave velocity along the z-axis.

Among localized waves, Bessel beams are particularly interesting solutions due to their self-healing property [3,19,20]. They are obtained from a spectrum equal to:

$$S(k_\rho, \omega) = \frac{\delta(k_\rho - k_0 \sin\theta_0)}{k_\rho}\delta(\omega - \omega_0). \tag{4}$$

where $\delta(\cdot)$ is the Dirac delta function. Equation (4) clearly implies that a Bessel beam is a monochromatic solution in $\omega_0 = 2\pi f_0$ (f_0 being the working frequency) with $k_\rho = k_0 \sin\theta_0 = (\omega_0/c_0)\sin\theta_0$, where $0 < \theta_0 < \pi/2$ is the so-called *axicon angle*, defined with respect to the vertical z-axis. Therefore, a Bessel beam can be represented as a superposition of plane waves with wavenumbers lying on the surface of a cone with aperture angle $2\theta_0$. From the separation relation, we also have $k_z = k_0 \cos\theta_0$, thus a linear coupling between ω and k_z is required for obtaining localized-wave solutions [18].

Using Equation (4) in (3), an ideal diffraction-free Bessel beam is obtained, i.e., a Bessel beam whose transverse intensity is independent from the propagating distance $|\psi(\rho, \phi, z > 0, t)|^2 = |\psi(\rho, \phi, z = 0, t)|^2$ [20]:

$$\psi(\rho, z, t) = J_0(k_\rho\rho)e^{-jk_z z}e^{j\omega_0 t} \tag{5}$$

As discussed in the Introduction, when generated from a finite aperture, this expression holds true only for $z \leq z_{\text{ndr}}$, where z_{ndr} is the so-called *nondiffractive range*. According to a ray-optics analysis, this distance is given by

$$z_{\text{ndr}} = \rho_{\text{ap}} \cot\theta_0 \tag{6}$$

where ρ_{ap} is the aperture radius of the radiating aperture, and for $z > z_{ndr}$ the field intensity abruptly decreases, i.e., $|\psi(\rho, z > z_{ndr}, t)|^2 \to 0$. Interestingly, the transverse spot size S_ρ of a Bessel beam is also related to the axicon angle through the following equation [21]:

$$S_\rho = 0.7655\lambda \csc\theta_0 \tag{7}$$

where λ is the vacuum wavelength. Clearly, z_{ndr} and S_ρ both increase as θ_0 decreases. In most applications, one is interested in having the largest z_{ndr} with the smallest ρ_{ap} and S_ρ, thus the choice of θ_0 is subject to the trade-off dictated by (6) and (7) [21].

3. Leaky-Wave Analysis

In this section, a step-by-step design workflow is provided and justified through a simple yet rigorous theoretical explanation. In Section 3.1, a leaky-wave approach is used in order to obtain the design equations for the resonant Bessel-beam launcher presented in Section 3.2. In Section 3.3, the dispersion analysis of the device is presented.

3.1. The Leaky-Wave Approach

Leaky-wave antennas (LWAs) belong to the broader class of traveling-wave antennas, for which the excitation is produced by a *leaky* wave that propagates along an open waveguide-like structure [22]. The main distinctive feature of LWAs is that the radiation mechanism can efficiently be described in terms of leaky waves, which are characteristic modes endowed with complex wavenumbers, capable of providing a compact yet accurate representation of the total aperture field [22]. As shown in [5,6], a Bessel beam can be generated by means of a two-dimensional (2D) leaky radial waveguide.

Such a device is typically constituted by a grounded dielectric slab encircled by a circular metallic rim with a PRS on top, as schematically shown in Figure 1. If the PRS is but a *slight* perturbation of a perfect electric conductor (PEC), the *leaky modes* propagating in the structure will be *slight* perturbations of the guided modes existing in the unperturbed equivalent parallel-plate waveguide. Under these conditions (*small* perturbations), the total aperture field is dominated by the leaky-wave contribution, which is sufficient to provide for an accurate description of the whole radiating mechanism. (A rigorous explanation and limitations to the application of the leaky-wave approach can be found in [23,24].)

In a 2D-LWA excited by a finite three-dimensional (3D) source, the field distribution is mainly described by cylindrical leaky waves that propagate in the transverse plane with a complex radial wavenumber $k_\rho = \beta_\rho - j\alpha_\rho$, where β_ρ and α_ρ are the phase and the leakage constant, respectively [25]. It is worthwhile recalling here that β_ρ is intimately related to the axicon angle θ_0 (which entirely defines the focusing properties of Bessel beams [21]) through the well-known ray-optics formula $\beta_\rho / k_0 = \sin\theta_0$ [26].

As is customary, the electromagnetic problem is studied separately for TM and TE waves. In such a way, only the z-component of the vector potential A_z or F_z is required to totally describe the TM or TE field, respectively. Regardless of the polarization type, resonant Bessel-beam launchers are typically excited by a dipole-like source: if a vertical electric dipole (VED) is used a TM field distribution is achieved, whereas a TE polarization is excited through a vertical magnetic dipole (VMD).

In the case of a TM- or TE-polarized, azimuthally symmetric device, the radial dependence of A_z or F_z, respectively, can be totally described in terms of two wave contributions: the *outward* cylindrical leaky wave excited by the dipole-like source, and the *inward* one coming from the reflection on the circular metallic rim. From an analytical viewpoint, by considering $H_0^{(1)}$ and $H_0^{(2)}$ the zeroth-order Hankel functions of the first and the second kind, respectively, it results in:

$$A_z(\rho, z) \propto A_0^+ H_0^{(2)}(k_\rho \rho) + A_0^- H_0^{(1)}(k_\rho \rho) \tag{8}$$

$$F_z(\rho, z) \propto F_0^+ H_0^{(2)}(k_\rho \rho) + F_0^- H_0^{(1)}(k_\rho \rho) \tag{9}$$

with $A_0^{\pm}$ ($F_0^{\pm}$) the complex amplitude coefficients of the outward/inward cylindrical leaky waves constituting the TM (TE) potentials. Now that we have defined the radial dependence, it remains to define the longitudinal dependence of A_z, which can simply be assumed as an outward travelling plane wave $e^{-jk_z z}$. The electric and magnetic field components on the aperture plane ($z = 0$) are then found from Maxwell's equations and, for a TM-polarized field, read:

$$\begin{cases} E_\rho(\rho) & = -\frac{k_z}{\omega\epsilon}\frac{\partial}{\partial\rho}A_z(\rho) \\ E_\phi(\rho) & = 0 \\ E_z(\rho) & = -j\frac{k_\rho^2}{\omega\epsilon}A_z(\rho) \end{cases} \begin{cases} H_\rho(\rho) & = 0 \\ H_\phi(\rho) & = -\frac{1}{\rho}\frac{\partial A_z(\rho)}{\partial\rho} \\ H_z & = 0 \end{cases} \tag{10}$$

For the TE-polarized case, the duality principle can be applied, which yields null E_ρ, E_z, and H_ϕ components, with E_ϕ and H_ρ proportional to $\frac{\partial}{\partial\rho}F_z$; H_z follows the same radial envelope of F_z.

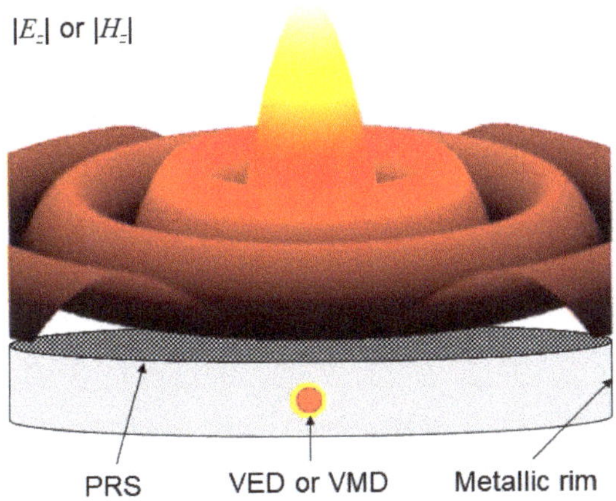

Figure 1. Pictorial representation of a TM- or TE-polarized resonant Bessel-beam launcher excited by a VED or a VMD with its $|E_z|$ or $|H_z|$ near-field distribution, respectively.

In order to obtain a Bessel beam, the aperture radius has to be properly set in such a way to have a constructive interference between the *outward* and *inward* cylindrical leaky waves that fully describe the potentials TM and TE potentials A_z and F_z. If we assume that the coefficient of the outward wave A_0^+ (F_0^+) is almost equal to that of the inward wave A_0^- (F_0^-), a $J_0(k_\rho\rho)$ distribution for A_z (F_z) is achieved for TM (TE) polarization with an amplitude $A_0 \equiv A_0^+ \simeq A_0^-$ ($F_0 \equiv F_0^+ \simeq F_0^-$). As a result, the vertical field distribution E_z (H_z) will also be characterized by a zeroth-order Bessel function. Therefore, non-null components for the TM- and TE- polarized case read:

$$\begin{cases} E_z(\rho,z) & = -jA_0\frac{k_\rho^2}{\omega\epsilon}J_0(k_\rho\rho)e^{-jk_z z} \\ E_\rho(\rho,z) & = A_0\frac{k_z k_\rho}{\omega\epsilon}J_1(k_\rho\rho)e^{-jk_z z} \\ H_\phi(\rho,z) & = A_0\frac{k_\rho}{\rho}J_1(k_\rho\rho)e^{-jk_z z} \end{cases} \tag{11}$$

and

$$\begin{cases} H_z(\rho,z) & = -jF_0\frac{k_\rho^2}{\omega\epsilon}J_0(k_\rho\rho)e^{-jk_z z} \\ H_\rho(\rho,z) & = F_0\frac{k_\rho k_z}{\omega\mu}J_1(k_\rho\rho)e^{-jk_z z} \\ E_\phi(\rho,z) & = -F_0 k_\rho J_1(k_\rho\rho)e^{-jk_z z} \end{cases}, \tag{12}$$

respectively. Design criteria to enforce both the radial and the vertical resonance will be explained in the following subsection.

3.2. Design Parameters

Due to the presence of the metallic rim, a radial resonance is found when the aperture radius coincides with one of the zeroes of the Bessel functions characterizing the radial dependence of the tangential electric fields. Therefore, different resonant conditions have to be applied depending on whether we deal with TM- or TE-polarized Bessel beams. In the former case, the tangential electric-field component is E_z that goes as $J_0(\cdot)$ (see (11)), whereas, in the latter case, it is E_ϕ that goes as $J_1(\cdot)$ (see (12)). Therefore, the aperture radius should satisfy the following equation:

$$\beta_\rho \rho_{\rm ap} = j_{nq}, \tag{13}$$

where j_{nq} is the qth zero of the nth order Bessel function, with $n = 0$ in the TM case (as shown in, e.g., [6,10]) and $n = 1$ in the TE case (as shown, e.g., in [13]).

From (6) and (13), a useful relation is achieved among the nondiffractive range, the aperture radius, the polarization type (which determines the order n of the Bessel function), and the order q of the radial resonance, which reads:

$$z_{\rm ndr} = \frac{\rho_{\rm ap}\sqrt{k_0^2\rho_{\rm ap}^2 - j_{nq}^2}}{j_{nq}} \tag{14}$$

In order to obtain a Bessel-beam distribution with a desired nondiffractive range and aperture radius (fixed by practical constraints), we show here a possible design workflow to determine the required value of β_ρ and α_ρ at the working frequency f_0 and the resulting design in terms of cavity height h and sheet reactance X_s; the latter is assumed to fully characterize a lossless PRS. This assumption is fairly accurate for canonical homogenized metasurfaces such as 2D patch arrays, metal strip gratings, and fishnet-like metasurfaces (see, e.g., [27,28]) as those discussed here. The synthesis of a fishnet-like metasurface, i.e., its practical implementation for realizing a given value of X_s, will be discussed in Section 4. Here, we first present design equations to properly set the relevant *design parameters* of a resonant Bessel-beam launcher, i.e., X_s, $\rho_{\rm ap}$, and h, for having a Bessel beam suitable for a given application scenario.

For this purpose, we assume the aperture radius and the nondiffractive range are fixed by some practical constraints dictated by a typical WPT scenario for wearable devices. In this context, it is commonly required to reach a minimum cover distance with a device that cannot exceed a given maximum transverse size [29], fixed in this case by the aperture radius. In this regard, Figure 2 shows through gray dashed lines how these boundaries would translate at 30, 60, and 90 GHz for a maximum aperture radius of 1 cm and a minimum cover distance of 1.5 cm. It is worthwhile to point out that, in Figure 2, the trade-off between miniaturization and cover distance for a given radial resonance is also clearly represented.

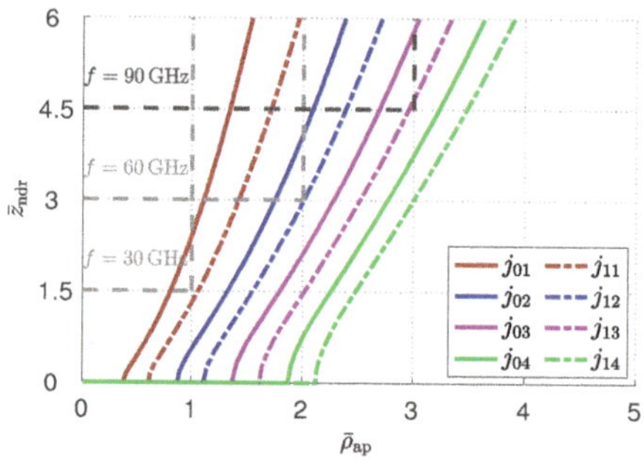

Figure 2. Design curves for a minimum $z_{\rm ndr} = 15$ mm and a maximum $\rho_{\rm ap} = 10$ mm. The normalized nondiffractive range $\bar{z}_{\rm ndr} = z_{\rm ndr}/\lambda_0$ and the normalized aperture radius $\bar{\rho}_{\rm ap} = \rho_{\rm ap}/\lambda_0$ are reported on the y-axis and the x-axis, respectively. Gray dashed lines represent the design boundaries at 30, 60, and 90 GHz, and differently colorized solid or dashed lines represent the design curves in a TM or TE polarization, respectively, for different radial resonances.

Once f_0 is fixed, the order q of the radial resonance and the polarization have to be chosen. In particular, the first, $q = 1$ radial resonance is typically avoided due to its consequent high truncation effects. Equation (13) is then used to get the normalized phase constant $\hat{\beta}_\rho = \beta_\rho / k_0$, strictly related to the axicon angle through the equation $\hat{\beta}_\rho = \sin\theta_0$, once the polarization ($n = 0$ or $n = 1$ for the TM or TE case, respectively) and the radial resonance q are fixed.

At this point, the normalized leakage constant $\hat{\alpha}_\rho = \alpha_\rho / k_0$ has to be determined. As noted before and shown in [6], we need $A_0^+ \simeq A_0^-$ ($F_0^+ \simeq F_0^-$). Under the simplified assumption of plane-wave propagation (which ignores the radial spreading of cylindrical leaky waves) inside the cavity, the ratio $0 < a_\mathrm{r} < 1$ between the power amplitudes of the inward and outward wave is given by $a_\mathrm{r} = e^{-2\alpha_\rho \rho_\mathrm{ap}}$. In order to obtain a well-defined Bessel-beam profile without requiring the design of a very high-Q (Q stands for *quality factor* [12]) cavity, the power ratio between the waves should be moderately close to 1. On a quantitative basis, this trade-off is substantially met for $a_\mathrm{r} \leq 0.95$, thus providing the following design equation for the leakage constant:

$$\hat{\alpha}_\rho < \frac{0.03}{k_0 \rho_\mathrm{ap}} \tag{15}$$

From the knowledge of the required $\hat{\beta}_\rho$ and $\hat{\alpha}_\rho$, the design equations for the cavity height h and the equivalent PRS reactance X_s can be found for either TE- and TM- polarized Bessel-beam launchers, using one of the methods explained in [30]. For the readers' convenience, we report here the approximate analytical equations for the reactance sheet X_s values:

$$|X_\mathrm{s}^\mathrm{TE}| = \eta_0 \sqrt{\frac{\pi \hat{\beta}_\rho \hat{\alpha}_\rho \sec\theta_0}{\sqrt{(\varepsilon_\mathrm{r} - \sin^2\theta_0)^3}}} \tag{16a}$$

$$|X_\mathrm{s}^\mathrm{TM}| = \eta_0 \sqrt{\frac{\pi \hat{\beta}_\rho \hat{\alpha}_\rho \cos\theta_0}{\varepsilon_\mathrm{r}\sqrt{\varepsilon_\mathrm{r} - \sin^2\theta_0}}}. \tag{16b}$$

and cavity height h values:

$$h_\mathrm{TE} \simeq \frac{\lambda_0}{2\sqrt{\varepsilon_\mathrm{r} - \sin^2\theta_0}} \left(1 - \frac{X_\mathrm{s}\sqrt{\varepsilon_\mathrm{r} - \sin^2\theta_0}}{\pi\eta_0}\right) \tag{17a}$$

$$h_\mathrm{TM} \simeq \frac{\lambda_0}{2\sqrt{\varepsilon_\mathrm{r} - \sin^2\theta_0}} \left(1 - \frac{X_\mathrm{s}\varepsilon_\mathrm{r}}{\pi\eta_0\sqrt{\varepsilon_\mathrm{r} - \sin^2\theta_0}}\right). \tag{17b}$$

where $\eta_0 \simeq 120\pi\ \Omega$ is the vacuum impedance. The absolute value in (16a) and (16b) is used because the design equations apply for both inductive $X_\mathrm{s} > 0$ and capacitive $X_\mathrm{s} < 0$ sheets. We also note that Bessel-beam launchers typically work far from the leaky cutoff condition (viz., $\hat{\beta}_\rho \simeq \hat{\alpha}_\rho$ [31]) and usually require highly reflective PRS (which leads to small values of $\hat{\alpha}_\rho$). Therefore, the accuracy of the aforementioned equations is rather high [30]. The effectiveness of the entire design workflow can be inferred from the dispersion curves of the TE and TM leaky modes, here obtained with an accurate numerical routine, briefly discussed in the next subsection.

3.3. Dispersion Analysis

The propagation modes supported by the leaky-wave resonant cavity can be found by enforcing the condition of resonance on the relevant transverse equivalent network (TEN), thus obtaining the modal dispersion equation. The dispersion equation can be obtained by enforcing the resonance condition on the relevant TEN. In other words, we apply the so-called transverse resonance technique [32] to the TEN represented in Figure 3a where Y_0 and Y_1 are the characteristic admittances in air and inside the cavity, respectively, $Y_\mathrm{s} = 1/Z_\mathrm{s}$ is the sheet admittance, and k_z and k_{z1} are the vertical wavenumber in air and inside the cavity, respectively. While the vertical wavenumbers are again given by $k_z = \sqrt{k_0^2 - k_\rho^2}$ and $k_{z1} = \sqrt{k_0^2 \varepsilon_r - k_\rho^2}$, the characteristic admittances have the well-known different expressions reported in Table 1 [12], depending on the specific polarization (viz., TE or TM) and material.

Table 1. Expression for the characteristic admittances in different material and polarizations.

Admittance	TE	TM
Y_0	$\frac{k_z}{k_0 \eta_0}$	$\frac{k_0}{k_z \eta_0}$
Y_1	$\frac{k_{z1}}{k_0 \eta_0}$	$\frac{k_0 \varepsilon_r}{k_{z1} \eta_0}$

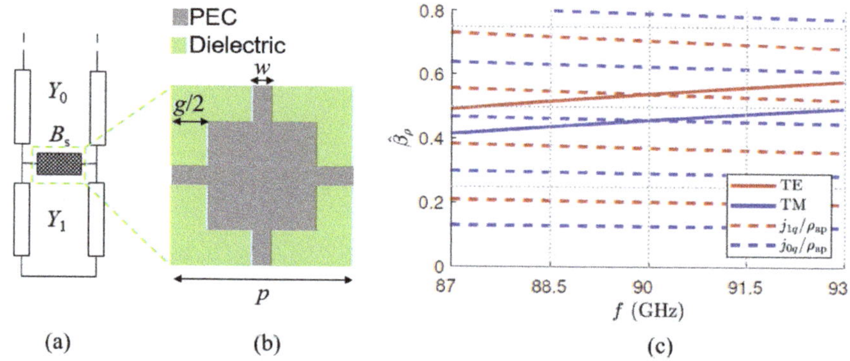

Figure 3. (**a**) Transverse equivalent network of a resonant Bessel-beam launcher, and (**b**) the geometry of the PRS fishnet-like unit cell. The areas in which PEC is not present are filled by the dielectric inside the cavity, constituted by air in this case. (**c**) Dispersion diagrams of the normalized phase constant $\hat{\beta}_\rho$ vs. frequency f for the specific case derived in Section 4, with the design parameters reported in Table 2. Red and blue solid lines represent the dispersion curves of TE and TM leaky modes, respectively. The radial resonances for TE and TM modes, represented by red and blue dashed lines, are obtained by enforcing a null of the first- or the zeroth-order Bessel function, respectively.

Table 2. Design parameters and radial wavenumbers for TE- and TM-polarized resonant Bessel-beam launchers with a working frequency $f_0 = 90$ GHz and an aperture radius $\rho_{ap} = 10$ mm. The period of the PRS unit cell is $p \simeq 3.33$ mm.

Polarization	$\hat{\beta}_\rho$	$\hat{\alpha}_\rho$	X_s (Ω)	h (mm)	g/p	w/p
TE	0.5401	0.0019	30	1.94	0.2086	0.05
TM	0.4588	0.0020	20	1.84	0.1732	0.1

Therefore, we obtain the following dispersion equation:

$$Y_0(k_\rho) + jB_s - jY_1(k_\rho) \cot(k_{z1}(k_\rho)h) = 0, \quad (18)$$

where $B_s = -1/X_s$, and the dependence of all terms on k_ρ is explicit (we are tacitly assuming the PRS to be not spatially dispersive, and polarization dependent; a fair approximation for fishnet-like metasurfaces [28]). The radial wavenumbers k_ρ of all the modes propagating in the structure are found by solving for the complex roots of (18). It should be noted that the left-hand side of (18) is a two-sheeted Riemann surface, whose complex planes are connected through the Sommerfeld branch cuts. Leaky modes are found searching for the *complex improper roots*, lying on the *improper* sheet, i.e., with Im $(k_z) > 0$. The dispersion diagrams, i.e., $\hat{\beta}_\rho$ or $\hat{\alpha}_\rho$ as functions of the frequency, are then obtained and commented for the case study analyzed in Section 4.

4. Structure Design

Here, we exploit the theoretical results of Section 3 to present a realistic implementation for both TE- and TM-polarized resonant Bessel-beam launchers.

The first step in the implementation of a prototype is fixing some parameters. First of all, we can assume that our device will work in a realistic scenario for WPT where it has to respect specific dimension constraints. For instance, in [33], a maximum aperture radius of 1.5 cm and a nondiffractive range of 2 cm at 30 GHz were required. In this work, a further miniaturization of the device is set by fixing $\rho_{\mathrm{ap}} = 1$ cm, at the expense of a slightly lower nondiffractive range of $z_{\mathrm{ndr}} = 1.5$ cm (note that the ratio $z_{\mathrm{ndr}}/\rho_{\mathrm{ap}}$ is increased). These boundaries are reported in gray dashed lines in Figure 2, where $\bar{z}_{\mathrm{ndr}}$ and $\bar{\rho}_{\mathrm{ap}}$ are the normalized nondiffractive range and aperture radius with respect to the vacuum wavelength, respectively. From Figure 2, we note that, for $f_0 = 30$ GHz, we can only use the first-order radial resonance that is preferable to avoid, as it may lead to non-negligible truncation effects. Conversely, for $f_0 = 60$ GHz and $f_0 = 90$ GHz, we can choose up to the second $q = 2$ and third $q = 3$ radial resonance, respectively. In particular, the latter has been considered in this work as a practical case study.

Once the radial resonance, the aperture radius, and the power ratio $a_{\mathrm{r}} = 0.95$ are fixed, the radial wavenumber k_ρ is obtained, as previously described. With these values of β_ρ and α_ρ at hand, the equivalent PRS surface impedance Z_{s} and the cavity height h are derived through (16a) and (16b) and (17a) and (17b), respectively. All these parameters are reported for brevity in Table 2. The sought value of Z_{s} is obtained through the fishnet-like unit cell shown in Figure 3b and commented in more detail in Section 4.

The theoretical effectiveness of the described design workflow is verified in Figure 3c. There, the dispersion curves of the TE and TM leaky-wave modes are reported vs. frequency f along with their different radial resonances. As expected, at the working frequency $f_0 = 90$ GHz, the phase-constant curves (solid lines) and the radial-resonance curves (dashed lines) intersect, thus defining the working point of the designed Bessel-beam launcher. For the TE and TM case, we found $\hat{\beta}_\rho = 0.5401$ and $\hat{\beta}_\rho = 0.4588$ (see Table 2), to which correspond $\theta_0 \simeq 32.7^\circ$ and $\theta_0 \simeq 27.3^\circ$ and in turn $S_\rho \simeq 4.72$ mm and $S_\rho \simeq 5.56$ mm, respectively.

The last steps of a the practical implementation for resonant Bessel-beam launcher are the physical realizations of a homogenized PRS with the desired equivalent X_{s} and of a realistic source that acts as a VMD or a VED.

4.1. PRS Implementation

A PRS is a key component for different kinds of antennas: uniform 2D leaky-wave antennas, Fabry–Perot cavity antennas, reconfigurable devices and, of course, resonant Bessel-beam launchers. PRS realizations mainly fall into three different categories: single-layer dielectric covers, dielectric multilayers, and homogenized metasurfaces [34]. In order to have a low-profile antenna and to avoid high-permittivity materials that usually exhibit high dielectric losses, it is preferred to use a metal homogenized metasurface for the implementation of the PRS in Bessel-beam launchers at millimeter waves. This solution also has the advantage of being relatively low cost due to its easy realization through standard PCB technologies or, at high frequencies, standard photolithographic processes. The PRS topology, given by square or circular patches, printed dipoles, or fishnet-like metasurfaces, depends on the type of the PRS (i.e., inductive or capacitive PRS) and the desired value of the equivalent impedance sheet.

The implementation of an inductive PRS with a relatively low X_{s} value can be achieved through a fishnet-like metasurface whose unit cell is reported in Figure 3b. The equivalent surface impedance can be tuned using different values for the patch gap g and the strip width w. As shown in [28], as the width increases or the gap decreases, the equivalent reactance X_{s} decreases and tends to represent a PEC, as expected.

Through a parametric full-wave analysis of the unit cell, the design parameters of the fishnet-like metasurface needed to realize the desired X_{s} in the TE and TM polarizations were achieved. These g and w values are reported normalized with respect to the period $p = \lambda/10 \simeq 3.33$ mm of the PRS unit cell in Table 2.

4.2. Feeder

So far, we have considered an ideal source given by a VED or a VMD for a TM- or a TE-polarized Bessel-beam launcher, respectively. This kind of source, however, is not a realistic feeder, thus it is necessary to take into account its physical implementation.

As concerns the VED, a low-cost, effective, simple solution is given by coaxial probes. In fact, coaxial probes behave as VED-like sources and are commercially available up to 110 GHz. Conversely, realizations of VMDs present different issues. In [13,14], loop antennas or coils were considered for exciting a TE-polarized Bessel beam. However, the feeding point of this kind of source necessarily breaks the azimuthal symmetry of the structure, deteriorating the TE polarization purity. The solution proposed in [13] that partially solves this issue is given by a specific shift of the probe position from the center obtained through a full-wave parametric optimization. In such a way, a zeroth-order Bessel-beam profile is achieved for the vertical component of the magnetic field, but spurious contributions are still present. It was recently shown [17] that it is possible to excite a purely TE-polarized Bessel-beam launcher by means of a radial, simultaneously excited slot array on the ground plane. Each slot can be excited by a microstrip, but this may need a rather complex feeding network, because the number of ports increases with the number of slots.

An innovative single-port excitation for a TE-polarized Bessel-beam launcher is then presented here. Whereas a radial slot array is the discrete counterpart of a radially directed magnetic surface current, we can excite this source in its continuous form by means of a TE_{01}-polarized circular waveguide. The TE_{01} mode is a higher-order mode in a circular waveguide, and it can be isolated through different implementations described in the available literature [35–37]. In order to reduce the computational burden, we have considered on CST Microwave Studio [38] an ideal waveguide port where only the TE_{01} mode is excited by ensuring, through the guide dimension, that the desired mode is above cutoff.

4.3. Comparison between Theoretical and Full-Wave Results

The consistency of the proposed approach is verified through a comparison between full-wave results and the numerical evaluation of expected theoretical field distributions according to the outlined leaky-wave approach. The former are obtained through a CST simulation of the entire three-dimensional model of the structure for both polarizations; the latter are obtained through the application of a Gauss–Legendre quadrature rule to the radiation integral of the theoretical aperture fields.

In particular, for the TM-polarized case, the aperture field distribution is given by (11) in $z = 0$. From the knowledge of the field, the equivalent magnetic and electric surface currents over the aperture plane are derived and used in the radiation integral to get the near-field distribution for $z \geq 0$ [39]. The typical profile of a zeroth-order Bessel beam is obtained for the longitudinal component of the electric field E_z, as shown in Figure 4, where the field amplitude over the ρz-plane is reported in dB, normalized with respect to its maximum, for both numerical and full-wave results. In both cases, a Bessel beam with a transverse spot size of about half a centimeter is maintained over a distance of 1.5 cm. An impressive agreement between theoretical and full-wave results is obtained, thus corroborating the effectiveness of the presented design workflow. In Figure 4, we can appreciate the typical on-axis oscillations of truncated Bessel beams. To circumvent this issue, and thus generate a quasi-homogeneous axial intensity distribution, apodization techniques (see, e.g., [40,41]) are often proposed in optics, as well as at microwave frequencies (see, e.g., [42,43]).

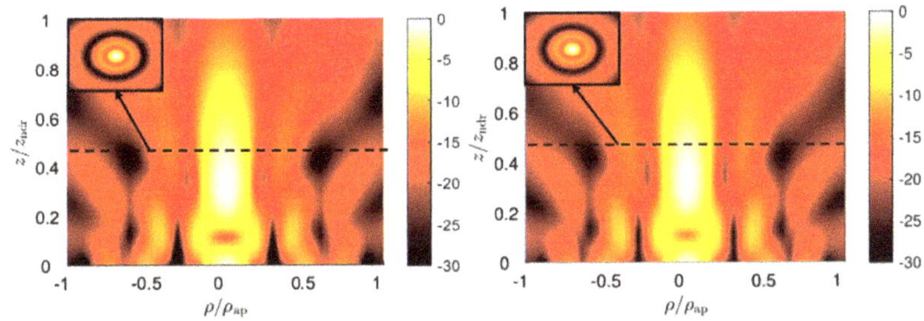

Figure 4. Full-wave and radiation-integral results for the longitudinal electric-field component E_z in the TM-polarized Bessel-beam launcher on the left and on the right, respectively. At the top-left corner, the $|E_z|$ field distribution is represented over the transverse xy-plane at $z = z_{\mathrm{ndr}}/2$.

As a consequence of the duality principle, similar results are obtained for the TE-polarized case shown in Figure 5, where the longitudinal component of the magnetic field $|H_z|$ is represented as per $|E_z|$ in Figure 4. In this case, the theoretical approach is the same used for the TM case, provided the aperture field distribution is now given by (12) [39].

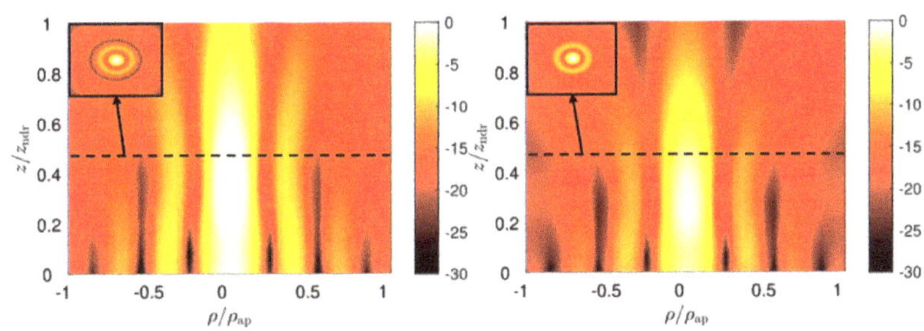

Figure 5. Full-wave and radiation-integral results for the longitudinal magnetic-field component H_z in the designed TE-polarized Bessel-beam launcher on the left and on the right, respectively. At the top-left corner, the $|H_z|$ field distribution is represented over the transverse xy-plane at $z = z_{\mathrm{ndr}}/2$.

5. Applications

In the previous sections, the theoretical implementation and the design workflow of resonant Bessel-beam launchers based on the leaky-wave approach were described and validated through full-wave simulations. Here, we show different practical scenarios that may take great advantage of such launchers.

Bessel beams, due to their remarkable features of beam focusing and self-healing in the near field, are very useful in diverse engineering fields [44]. For instance, high data-rate links or micromanipulation of small particles can be achieved through Bessel beams, as shown in [45,46], respectively. Bessel beams are also used in tomography, imaging [47–49], and ground penetrating radar applications [50,51], also with interesting reconfigurable properties [52]. Moreover, Bessel beams have attracted much interest in the THz range [53] for improving the depth-of-field [54]. They can be implemented through microstructured photoconductive antennas [55], plasmonic waves [56], or irregular binary axicons [57]. Interestingly, in the 3–300 GHz frequency range, Bessel-beam launchers present a very low-cost and simple implementation and a higher penetration depth with respect to their optical counterparts.

In this regard, WPT is one of the application scenarios where the generation of Bessel beams at 3–300 GHz has attracted much interest in the last decade. In particular, the Bessel-beam limited-diffractive behavior can be exploited in order to realize a radiative near-field WPT link. The radiative near-field link presents a higher efficiency with respect to far-field WPT links [58] and a higher covered distance with respect to the typical reactive near-field links based on inductive coupling [59]. As explained in [60], for distances beyond the reactive near field, magneto-inductive coupling becomes insignificant due to the decay of the link efficiency as the sixth power of the distance between transmitter and receiver.

Bessel beams have recently been studied for the implementation of radiative near-field WPT links [16,37]. In practical cases, such as the realization of wearable devices, it is very important to consider a compact structure for realizing a WPT link over a few wavelengths distance. For instance, in [17,29,33], a miniaturized radiative near-field link was established between two resonant Bessel-beam launchers. In such a way, a direct link between these two resonant devices was achieved, such as the one represented in Figure 6.

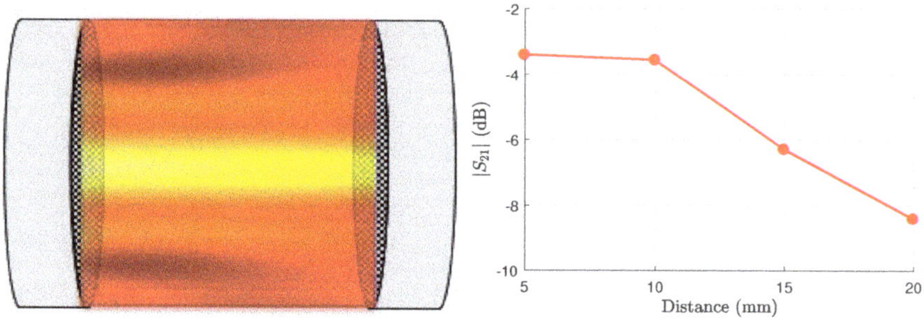

(**a**) (**b**)

Figure 6. (**a**) Pictorial representation of a radiative near-field WPT link between two resonant Bessel-beam launchers. (**b**) The $|S_{21}|$ values in a radiative near-field WPT link at 90 GHz for different distances between transmitting and receiving TM-polarized Bessel-beam launchers.

By exploiting the TM-polarized Bessel-beam design presented in this work, an innovative WPT link at 90 GHz can be realized. As shown in Figure 6a, a near-field radiative WPT link is realized between two identically polarized resonant Bessel-beam launchers placed one in front of the other. By considering different distances between such devices and assuming the link as a two-port network, the resulting $|S_{21}|$ is reported in Figure 6b. In particular, the link has been simulated on CST Microwave Studio at 90 GHz for different distances between the devices (5, 10, 15, and 20 mm), obtaining interesting and promising results for high-frequencies WPT links.

Moreover, Bessel beams can show very different features depending on the specific application. If the cover distance and the size of the launcher are the main concerns, Bessel beams can be realized to work with a very low axicon angle. This, however, comes at the expense of the transverse size of the beam waist, which increases for lower axicon angles [21]. Conversely, one is keen to use a higher axicon angle when the transverse resolution is the main concern and the cover distance is not very important. On the other hand, if it is required to maintain a narrow beam waist for large distances, the size of the launcher has to be increased. In such conditions, resonant Bessel-beam launchers are not the best choice, and it is preferred to resort to the wideband type (see, e.g., [7,61–63]) mentioned in the Introduction.

In any case, the limited-diffraction and self-healing character of Bessel beams will always represent an added value with respect to more common realizations for applications where it is important to avoid waste of power and to ensure coverage in non-line-of-sight scenarios.

6. Conclusions

In this work, we started from a very general theoretical framework to analyze Bessel beams as ideal nondiffractive solutions to the wave equation. Their practical generation at millimeter waves through planar devices commonly known as resonant Bessel-beam launchers was presented and explained under the frame of a rigorous leaky-wave analysis.

The latter has allowed for deriving simple, yet accurate analytical equations for developing a useful design workflow for such devices.

The effectiveness of such design guidelines were corroborated through full-wave simulations of a device of only 2 cm of transverse size capable of generating Bessel beams at 90 GHz that maintain a transverse spot size less than half a centimeter over a 1.5 cm distance.

The potentialities of the proposed device were finally tested in a realistic scenario, where two identical resonant launchers were placed one in front of the other to create an innovative wireless link at 90 GHz in the radiative near-field. The performance of such a link opens interesting perspectives for future applications of these devices in wireless-power-transfer and other focusing applications.

Author Contributions: Conceptualization W.F., P.B. and A.G.; methodology, E.N. and W.F.; software, E.N.; data processing, E.N.; writing—original draft preparation, E.N. and W.F.; writing—review and editing, all authors; supervision, P.B. and A.G. All authors have read and agreed to the published version of the manuscript.

Funding: This research was funded by Ministero dell'Università e della Ricerca (PRIN n. 2017YJE9XK).

Data Availability Statement: Not applicable.

Conflicts of Interest: The authors declare no conflict of interest.

Abbreviations

The following abbreviations are used in this manuscript:

LWA	Leaky-Wave Antenna
PCB	Printed-Circuit Board
PEC	Perfect Electric Conductor
PRS	Partially Reflecting Sheet
TE	Transverse Electric
TM	Transverse Magnetic
VED	Vertical Electric Dipole
VMD	Vertical Magnetic Dipole
WPT	Wireless Power Transfer

References

1. Hernández-Figueroa, H.E.; Zamboni-Rached, M.; Recami, E. *Localized Waves*; John Wiley & Sons: Hoboken, NJ, USA, 2007.
2. Hernández-Figueroa, H.E.; Zamboni-Rached, M.; Recami, E. *Nondiffracting Waves*; John Wiley & Sons: Weinheim, Germany, 2013.
3. McGloin, D.; Dholakia, K. Bessel beams: Diffraction in a new light. *Contemporary Phys.* **2005**, *46*, 15–28. [CrossRef]
4. Ettorre, M.; Pavone, S.C.; Casaletti, M.; Albani, M.; Mazzinghi, A.; Freni, A. Near-field focusing by non-diffracting Bessel beams. In *Aperture Antennas for Millimeter and Sub-Millimeter Wave Applications*; Springer: Cham, Switzerland, 2018; pp. 243–288.
5. Ettorre, M.; Grbic, A. Generation of propagating Bessel beams using leaky-wave modes. *IEEE Trans. Antennas Propag.* **2012**, *60*, 3605–3613. [CrossRef]
6. Fuscaldo, W.; Valerio, G.; Galli, A.; Sauleau, R.; Grbic, A.; Ettorre, M. Higher-order leaky-mode Bessel-beam launcher. *IEEE Trans. Antennas Propag.* **2016**, *64*, 904–913. [CrossRef]
7. Comite, D.; Fuscaldo, W.; Podilchak, S.K.; Hílario-Re, P.D.; Gómez-Guillamón Buendía, V.; Burghignoli, P.; Baccarelli, P.; Galli, A. Radially periodic leaky-wave antenna for Bessel beam generation over a wide-frequency range. *IEEE Trans. Antennas Propag.* **2018**, *66*, 2828–2843. [CrossRef]
8. Chiotellis, N.; Mendez, V.; Rudolph, S.M.; Grbic, A. Experimental demonstration of highly localized pulses (X waves) at microwave frequencies. *Phys. Rev. B* **2018**, *97*, 085136. [CrossRef]
9. Chiotellis, N.; Zhang, S.; Vardaxoglou, Y.C.; Grbic, A. X wave radiator implemented with 3D printed metamaterials. *IEEE Trans. Antennas Propag.* **2020**, *68*, 5478–5486. [CrossRef]
10. Ettorre, M.; Rudolph, S.M.; Grbic, A. Generation of propagating Bessel beams using leaky-wave modes: Experimental validation. *IEEE Trans. Antennas Propag.* **2012**, *60*, 2645–2653. [CrossRef]
11. Fuscaldo, W.; Burghignoli, P.; Galli, A. Genealogy of leaky, surface, and plasmonic modes in partially open waveguides. *Phys. Rev. Appl.* **2022**, *17*, 034038. [CrossRef]
12. Pozar, D.M. *Microwave Engineering*; John Wiley & Sons: Hoboken, NJ, USA, 2009.

13. Lu, P.; Voyer, D.; Bréard, A.; Huillery, J.; Allard, B.; Lin-Shi, X.; Yang, X.S. Design of TE-polarized Bessel antenna in microwave range using leaky-wave modes. *IEEE Trans. Antennas Propag.* **2017**, *66*, 32–41. [CrossRef]
14. Lu, P.; Bréard, A.; Huillery, J.; Yang, X.S.; Voyer, D. Feeding coils design for TE-polarized Bessel antenna to generate rotationally symmetric magnetic field distribution. *IEEE Antennas Wirel. Propag. Lett.* **2018**, *17*, 2424–2428. [CrossRef]
15. Negri, E.; Fuscaldo, W.; Ettorre, M.; Burghignoli, P.; Galli, A. Analysis of resonant Bessel-beam launchers based on isotropic metasurfaces. In Proceedings of the 16th European Conference on Antennas and Propagation (EuCAP 2022), Madrid, Spain, 27 March–1 April 2022; pp. 1–4.
16. Heebl, J.D.; Ettorre, M.; Grbic, A. Wireless links in the radiative near field via Bessel beams. *Phys. Rev. Appl.* **2016**, *6*, 034018. [CrossRef]
17. Negri, E.; Benassi, F.; Fuscaldo, W.; Masotti, D.; Burghignoli, P.; Costanzo, A.; Galli, A. Effective TE-polarized Bessel-beam excitation for wireless power transfer near-field links. In Proceedings of the 52nd European Microwave Conference (EuMC), Milan, Italy, 25–30 September 2022; pp. 1–4.
18. Zamboni-Rached, M.; Recami, E.; Hernández-Figueroa, H.E. New localized superluminal solutions to the wave equations with finite total energies and arbitrary frequencies. *Eur. Phys. J. D* **2002**, *21*, 217–228. [CrossRef]
19. Durnin, J.; Miceli, J.J., Jr.; Eberly, J.H. Diffraction-free beams. *Phys. Rev. Lett.* **1987**, *58*, 1499. [CrossRef]
20. Durnin, J. Exact solutions for nondiffracting beams. I. The scalar theory. *J. Opt. Soc. Am. A* **1987**, *4*, 651–654. [CrossRef]
21. Fuscaldo, W.; Pavone, S.C. Metrics for localized beams and pulses. *IEEE Trans. Antennas Propag.* **2020**, *68*, 1176–1180. [CrossRef]
22. Galli, A.; Baccarelli, P.; Burghignoli, P. Leaky-Wave Antennas. *Wiley Enc. Electric. Electron. Eng.* **2016**, *2016*, 1–20.
23. Tamir, T.; Oliner, A.A. Guided complex waves. Part 1: Fields at an interface. *Proc. IEE* **1963**, *110*, 310–324. [CrossRef]
24. Tamir, T.; Oliner, A.A. Guided complex waves. Part 2: Relation to radiation patterns. *Proc. IEE* **1963**, *110*, 325–334. [CrossRef]
25. Burghignoli, P.; Fuscaldo, W.; Comite, D.; Baccarelli, P.; Galli, A. Higher-order cylindrical leaky waves–Part I: Canonical sources and radiation formulas. *IEEE Trans. Antennas Propag.* **2019**, *67*, 6735–6747 . [CrossRef]
26. Felsen, L. Real spectra, complex spectra, compact spectra. *J. Opt. Soc. Am. A* **1986**, *3*, 486–496. [CrossRef]
27. Luukkonen, O.; Simovski, C.; Granet, G.; Goussetis, G.; Lioubtchenko, D.; Raisanen, A.V.; Tretyakov, S.A. Simple and accurate analytical model of planar grids and high-impedance surfaces comprising metal strips or patches. *IEEE Trans. Antennas Propag.* **2008**, *56*, 1624–1632. [CrossRef]
28. Fuscaldo, W.; Tofani, S.; Zografopoulos, D.C.; Baccarelli, P.; Burghignoli, P.; Beccherelli, R.; Galli, A. Systematic design of THz leaky-wave antennas based on homogenized metasurfaces. *IEEE Trans. Antennas Propag.* **2018**, *66*, 1169–1178. [CrossRef]
29. Benassi, F.; Fuscaldo, W.; Masotti, D.; Galli, A.; Costanzo, A. Wireless power transfer in the radiative near-field through resonant Bessel-beam launchers at millimeter Waves. In Proceedings of the 2021 IEEE Wireless Power Transfer Conference (WPTC), San Diego, CA, USA, 1–4 June 2021; pp. 1–4.
30. Fuscaldo, W.; Galli, A.; Jackson, D.R. Optimization of 1-D unidirectional leaky-wave antennas based on partially reflecting sheets. *IEEE Trans. Antennas Propag.* **2022**, *70*, 7853–7868. [CrossRef]
31. Lovat, G.; Burghignoli, P.; Jackson, D.R. Fundamental properties and optimization of broadside radiation from uniform leaky-wave antennas. *IEEE Trans. Antennas Propag.* **2006**, *54*, 1442–1452. [CrossRef]
32. Sorrentino, R.; Mongiardo, M. Transverse resonance techniques. In *Encyclopedia of RF and Microwave Engineering*; John Wiley & Sons, Ltd: Hoboken, NJ, USA, 2005.
33. Benassi, F.; Fuscaldo, W.; Negri, E.; Paolini, G.; Augello, E.; Masotti, D.; Burghignoli, P.; Galli, A.; Costanzo, A. Comparison between hybrid- and TM-polarized Bessel-beam launchers for wireless power transfer in the radiative near-field at millimeter waves. In Proceedings of the 51st European Microwave Conference (EuMC 2021), London, UK, 4–6 April 2022; pp. 1–4.
34. Burghignoli, P.; Fuscaldo, W.; Galli, A. Fabry–Perot cavity antennas: The leaky-wave perspective. *IEEE Antennas Propag. Mag.* **2021**, *63*, 116–145. [CrossRef]
35. Xu, Y.; Peng, T.; Sun, M.; Luo, Y.; Wang, J.; Jiang, W.; Liu, G.; Wu, Z. Design and test of broadband rectangular waveguide TE_{10} to circular waveguide TE_{21} and TE_{01} mode converters. *IEEE Trans. Electron Devices* **2019**, *66*, 3573–3579. [CrossRef]
36. Chu, Q.X.; Mo, D.Y.; Wu, Q.S. An isolated radial power divider via circular waveguide TE_{01}-mode transducer. *IEEE Trans. Microw. Theory Tech.* **2015**, *63*, 3988–3996. [CrossRef]
37. Paković, S.; Zhou, S.; González-Ovejero, D.; Pavone, S.C.; Grbic, A.; Ettorre, M. Bessel–Gauss beam launchers for wireless power transfer. *IEEE Open J. Antennas Propag.* **2021**, *2*, 654–663. [CrossRef]
38. CST Studio Suite. Electromagnetic Field Simulation Software, Dassault Systèmes. Available online: https://www.3ds.com/products-services/simulia/products/cst-studio-suite/ (accessed on 1 October 2022).
39. Balanis, C.A. *Advanced Engineering Electromagnetics*; Wiley: Hoboken, NJ, USA, 2012.
40. Ruschin, S. Modified Bessel nondiffracting beams. *J. Opt. Soc. Am. A* **1994**, *11*, 3224–3228. [CrossRef]
41. Stsepuro, N.; Nosov, P.; Galkin, M.; Krasin, G.; Kovalev, M.; Kudryashov, S. Generating Bessel-Gaussian beams with controlled axial intensity distribution. *Appl. Sci.* **2020**, *10*, 7911. [CrossRef]
42. Abielmona, S.; Gupta, S.; Caloz, C. Compressive receiver using a CRLH-based dispersive delay line for analog signal processing. *IEEE Trans. Microw. Theory Tech.* **2009**, *57*, 2617–2626. [CrossRef]
43. Fuscaldo, W.; Benedetti, A.; Comite, D.; Baccarelli, P.; Burghignoli, P.; Galli, A. Bessel-Gauss beams through leaky waves: Focusing and diffractive properties. *Phys. Rev. Appl.* **2020**, *13*, 064040. [CrossRef]

44. Pavone, S.C.; Sorbello, G.; Di Donato, L. Improving physical optics approximation through Bessel beam scattering. *IEEE Antennas Wirel. Propag. Lett.* **2021**, *20*, 993–997. [CrossRef]
45. Chen, S.; Li, S.; Zhao, Y.; Liu, J.; Zhu, L.; Wang, A.; Du, J.; Shen, L.; Wang, J. Demonstration of 20-Gbit/s high-speed Bessel beam encoding/decoding link with adaptive turbulence compensation. *Opt. Lett.* **2016**, *41*, 4680–4683. [CrossRef]
46. Arlt, J.; Garcés-Chávez, V.; Sibbett, W.; Dholakia, K. Optical micromanipulation using a Bessel light beam. *Opt. Commun.* **2001**, *197*, 239–245. [CrossRef]
47. Lee, K.S.; Rolland, J.P. Bessel beam spectral-domain high-resolution optical coherence tomography with micro-optic axicon providing extended focusing range. *Opt. Lett.* **2008**, *33*, 1696–1698. [CrossRef]
48. Planchon, T.A.; Gao, L.; Milkie, D.E.; Davidson, M.W.; Galbraith, J.A.; Galbraith, C.G.; Betzig, E. Rapid three-dimensional isotropic imaging of living cells using Bessel beam plane illumination. *Nat. Methods* **2011**, *8*, 417–423. [CrossRef]
49. Liu, C.; Zhao, Z.; Jin, C.; Xiao, Y.; Gao, G.; Xie, H.; Dai, Q.; Yin, H.; Kong, L. High-speed, multi-modal, label-free imaging of pathological slices with a Bessel beam. *Biomed. Opt. Express* **2020**, *11*, 2694–2704. [CrossRef]
50. Mugnai, D.; Spalla, P. Electromagnetic propagation of Bessel-like localized waves in the presence of absorbing media. *Opt. Commun.* **2009**, *282*, 4668–4671. [CrossRef]
51. Mazzinghi, A.; Balma, M.; Devona, D.; Guarnieri, G.; Mauriello, G.; Albani, M.; Freni, A. Large depth of field pseudo-Bessel beam generation with a RLSA antenna. *IEEE Trans. Antennas Propag.* **2014**, *62*, 3911–3919. [CrossRef]
52. Feng, R.; Ratni, B.; Yi, J.; Jiang, Z.; Zhang, H.; de Lustrac, A.; Burokur, S.N. Flexible manipulation of Bessel-like beams with a reconfigurable metasurface. *Adv. Opt. Mat.* **2020**, *8*, 2001084. [CrossRef]
53. Fuscaldo, W.; Pavone, S.C. Tunable terahertz Bessel-beam launchers. *Rev. Electromagn.* **2022**, *1*, 1–4. [CrossRef]
54. Bitman, A.; Moshe, I.; Zalevsky, Z. Improving depth-of field in broadband THz beams using nondiffractive Bessel beams. *Opt. Lett.* **2012**, *37*, 4164–4166. [CrossRef] [PubMed]
55. Winnerl, S.; Zimmermann, B.; Peter, F.; Schneider, H.; Helm, M. Terahertz Bessel-Gauss beams of radial and azimuthal polarization from microstructured photoconductive antennas. *Opt. Express* **2009**, *17*, 1571–1576. [CrossRef] [PubMed]
56. Monnai, Y.; Jahn, D.; Withayachumnankul, W.; Koch, M.; Shinoda, H. Terahertz plasmonic Bessel beamformer. *Appl. Phys. Lett.* **2015**, *106*, 021101. [CrossRef]
57. Yu, Y.Z.; Dou, W.B. Production of THz pseudo-Bessel beams with uniform axial intensity using irregular binary axicons. *IET Optoelectron.* **2010**, *4*, 195–200. [CrossRef]
58. Garnica, J.; Chinga, R.A.; Lin, J. Wireless power transmission: From far field to near field. *Proc. IEEE* **2013**, *101*, 1321–1331. [CrossRef]
59. Jolani, F.; Yu, Y.; Chen, Z. A planar magnetically coupled resonant wireless power transfer system using printed spiral coils. *IEEE Antennas Wirel. Propag. Lett.* **2014**, *13*, 1648–1651. [CrossRef]
60. Gowda, V.R.; Yurduseven, O.; Lipworth, G.; Zupan, T.; Reynolds, M.S.; Smith, D.R. Wireless power transfer in the radiative near field. *IEEE Antennas Wirel. Propag. Lett.* **2016**, *15*, 1865–1868. [CrossRef]
61. Fuscaldo, W.; Comite, D.; Boesso, A.; Beccarelli, P.; Burghignoli, P.; Galli, A. Focusing leaky waves: A class of electromagnetic localized waves with complex spectra. *Phys. Rev. Appl.* **2018**, *9*, 054005. [CrossRef]
62. Pavone, S.C.; Ettorre, M.; Casaletti, M.; Albani, M. Transverse circular-polarized Bessel beam generation by inward cylindrical aperture distribution. *Opt. Express* **2016**, *24*, 11103–11111. [CrossRef]
63. Pavone, S.C.; Ettorre, M.; Albani, M. Analysis and design of Bessel beam launchers: Longitudinal polarization. *IEEE Trans. Antennas Propag.* **2016**, *64*, 2311–2318. [CrossRef]

micromachines

MDPI

Article

Design and Fabrication of the Split Ring Resonator Shaped Two-Element MIMO Antenna with Multiple-Band Operation for WiMAX/5G/Zigbee/Wi-Fi Applications

Ammar Armghan [1,*], Khaled Aliqab [1], Vishal Sorathiya [2], Fayadh Alenezi [1], Meshari Alsharari [1] and Farman Ali [3]

[1] Department of Electrical Engineering, College of Engineering, Jouf University, Sakaka 72388, Saudi Arabia
[2] Faculty of Engineering and Technology, Parul Institute of Engineering and Technology, Parul University, Waghodia Road, Vadodara 391760, India
[3] Department of Electrical Engineering, Qurtuba University of Science and IT, Dera Ismail Khan 29050, Pakistan
* Correspondence: aarmghan@ju.edu.sa

Abstract: In this manuscript, we proposed the split ring resonator loaded multiple-input multiple-output (MIMO) antenna design for the frequency range of 1 and 25 GHz. The proposed antenna is numerically investigated and fabricated to analyze the different antenna parameters. We provided statistics on a wide range of antenna parameters for five different designs, including a simple circular patch antenna, a single-split-ring antenna, and a double-split-ring antenna. Reflectance, gain, directivity, efficiency, peak gain, and electric field distribution are all analyzed for all proposed antennas. The maximum achievable bandwidth is 5.28 GHz, and the double-split-ring resonator structure achieves this with a return loss of −20.84 dB. The radiation patterns of all the antenna with different port excitation conditions are presented to identify the behavior of the antenna radiation. We found the effect of the split-ring resonators to form radiation beams in different directions. We found the maximum and minimum half-power beam widths of 75° and 2°, respectively, among the different antenna designs. It was found that the split-ring resonator geometries in patch antenna convert wide-beam antenna radiation patterns to several narrow-beam radiation patterns. We found that each antenna's bandwidth, gain, and return loss performance significantly differs from the others. Overall, the proposed results of the antenna may apply to a wide range of communication applications, including those for Wi-Fi, WiMAX, and 5G.

Keywords: MIMO; 5G communication; antenna; gigahertz; gain; bandwidth; WiMAX; Wi-Fi

Citation: Armghan, A.; Aliqab, K.; Sorathiya, V.; Alenezi, F.; Alsharari, M.; Ali, F. Design and Fabrication of the Split Ring Resonator Shaped Two-Element MIMO Antenna with Multiple-Band Operation for WiMAX/5G/Zigbee/Wi-Fi Applications. *Micromachines* **2022**, *13*, 2161. https://doi.org/10.3390/mi13122161

Academic Editors: Trushit Upadhyay and Mark L. Adams

Received: 18 October 2022
Accepted: 3 December 2022
Published: 7 December 2022

Publisher's Note: MDPI stays neutral with regard to jurisdictional claims in published maps and institutional affiliations.

Copyright: © 2022 by the authors. Licensee MDPI, Basel, Switzerland. This article is an open access article distributed under the terms and conditions of the Creative Commons Attribution (CC BY) license (https://creativecommons.org/licenses/by/4.0/).

1. Introduction

The current prospective technologies help to try high data rates and low-priced applications. In line with their wideband systems can rapidly transfer a great deal of data across a wide variety of frequency channels while using a relatively low power spectral level [1]. There has been a tremendous shift in wireless communication networks over the past decade, yet the demand for compact, portable devices has remained steady. New wireless networks aim to accommodate many users simultaneously, providing variable data speeds and a versatile set of services. However, spectral congestion limits the available bandwidth, a barrier to increased data rates. Therefore, antenna solutions that allow integration across several frequency bands are becoming increasingly crucial for wireless devices [2]. Already in many smartphones and other portable devices, they may soon find broader application in consumer electronics [3]. Multi-input, multioutput (MIMO) technology is also employed to reduce multipath fading and expand available channel capacity [4,5]. MIMO enters the scene to increase system capacity and dependability. Multi-input multioutput (MIMO) relies on a large number of independent antennas for both transmission and reception. Different metrics can assess antenna array performance in MIMO systems [6,7].Using multiple-input multiple-output (MIMO) systems with multiple antenna modules at both the transmitter and receiver may improve wireless systems' multipath performance [8]. With MIMO technology, all antennas must be active at the same time.

The research shows that when the MIMO structure gets smaller, the isolation between the antenna components also decreases [9]. With the help of a radial stub-loaded resonator, MIMO antennas consisting of two antenna components have been constructed [10]. A flawed isolation wall is used to probe methods for decreasing mutual coupling, as presented in [11]. The results of the antenna presented in [12] demonstrate that isolation and bandwidth are sufficient, but they do not consider the antennas' diversity characteristics. Slotted triangular parasitic patches and two PIN diode switches [13,14] are used in MIMO to increase the principal triangle radiator's electrical length, enabling triple-band capabilities. When trying to achieve sub-6 GHz applications, the diagonally trunked patch with the partial diffracted ground concept is a useful tool, as depicted in the article [15]. Using an L-shaped slot in the radiating element and two identical stubs coupled to the partial ground improves the broadband radiation characteristics and impedance matching throughout the bands of interest [16]. In [17], a multi-input multioutput (MIMO) antenna based on four monopole antennas is proposed, enabling communication at mm-wave and sub-6 GHz frequencies while maintaining high isolation. An extensive and detailed pattern is conceivable because of the two layers of the mushroom structure. The inverted L-monopole construction [18] allows the four-element wideband MIMO antenna to cover an extensive frequency range. However, the achieved isolation is just −11 dB across the usable frequency range. There is much interference observed in the proposed antenna [19] because of the inadequate isolation between the antenna sections. When an antenna is constructed from a single element, the center frequency changes if a common ground is used and if the components are not well-isolated from one another [20]. In [21], the author proposes using a two-element MIMO antenna that has band-notch characteristics. The antenna is relatively large, but its strength is only 2 dBi. The two-element MIMO antenna presented in [22] has a wide operating frequency range of 2.1–11 GHz. On the other hand, we only managed to achieve a gain of 4 dBi and isolation of −22 dB. The machine learning capability of the MIMO antenna allows it to be used on a broader variety of settings. Federated learning (FL) is a machine learning form that considers privacy concerns by having users train models locally before sending their tweaks to a central server. Since data owners are not required to participate in FL, it is a privacy-invasive machine-learning approach [23,24]. In sum, these characteristics facilitate the realization of IoT-based applications. Interest in the next-generation 5G heterogeneous networks is widespread. There are huge benefits and many worries associated with using the bandwidth spectrum of HetNets to transfer massive amounts of data at rapid speeds [25]. Creating a reliable 5G antenna design is a key step toward maximizing the potential of the 5G network [26]. A transparent MIMO antenna made from a plexiglass base and transparent conductive oxide is described in the paper [27]. The proposed MIMO design for state-of-the-art 5G technology in [28] takes a novel method to increase isolation between several antenna components. Using two quasi-self-complimentary MIMO antennas to achieve broad bandwidths is proposed in [29]. This antenna is perfect because of its high gain and compact size. Due to its position, a lower ECC and lower isolation than other portions of the system are present in the quasi-structure. In the 2.3–2.4 GHz operating range, up to −40 dB isolation can be attained using the mushroom structure in a four-element MIMO antenna [30]. Massive MIMO is a frontrunner for the 5G network due to its great spectral efficiency and low power consumption. However, pilot contamination is a major barrier for huge MIMO systems (PC). Thus, a two-pilot scheduling approach is analogous to a synchronous fractional pilot scheduling system (AFPS) and a fractional pilot reuse (FPR) [31].

The proposed antennas in the previous literature survey provide a limited scope of the antennas' performance with specific gain and isolation properties. The bandwidth properties for the wide range of GHz applications are also limited in previously designed antennas. Therefore, it is essential to cover a wide range of GHz bands through the single-antenna design, which offers better isolation, wideband, and improvised gain. It is also required to design an antenna which can be worked on multiband applications for a wide range of device adaptability. Inspired by the previous antenna design and its application in wireless communication, we designed the MIMO antenna, which offers wideband operation with minimum return loss values and high radiation gain. We also designed an antenna that offers multiband operations and can work on narrow-band operation for specific antenna radiation applications. Overall, the antenna design proposed in this research can be applied in wideband, short-band, and multiband operations.

We simulated and fabricated a dual-element MIMO antenna structure for the 1 to 25 GHz operating frequency band. The proposed antenna is analyzed with different antenna patch structures, which include the circular patch, single split-ring-engraved and double split-ring resonator based on radiating elements. We investigated various antenna parameters such as return loss, gain, bandwidth, directivity, radiation pattern, efficiency, and normalized electric field to identify the overall behavior of the antenna. The proposed antenna can offer multiband resonating behavior for the 1 to 25 GHz of the frequency band.

2. Split Ring Resonator Loaded MIMO Antenna Structure

A split ring resonator based two-element MIMO antenna structure is numerical and experimentally analyzed over the 1 to 25 GHz frequency range frequency spectrum. The schematic of the proposed antenna is shown in Figure 1. Figure 1a shows the perspective view of the antenna. Figure 1b shows the top and bottom view of the antenna with the notation of the dimensions. The values of each dimension are shown in Table 1. The overall size of the antenna is set as 39 × 50 mm^2.

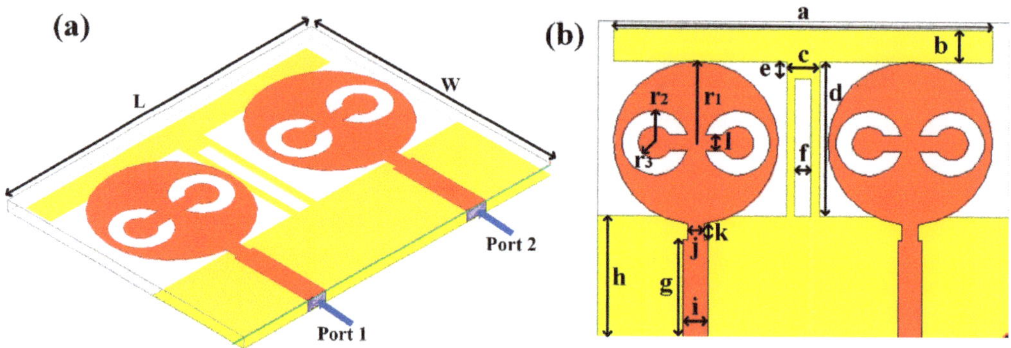

Figure 1. (**a**) Schematic of the split ring resonator loaded MIMO antenna. (**b**) The top (red) and bottom (yellow) layer structure is considered conductive copper material on FR4 substrate.

Table 1. Antenna design parameters.

Parameter	a	b	c	d	e	f	g	h	i	j	k	l	W	L	r_1	r_2	r_3
Values (mm)	46	4	4	19	2	2	12	15	3	2	2	3	39	50	10	4	2

We fabricated the five antenna structures to identify other design parameters and shapes. These antenna and its description are shown in Table 2. These antennas have two circular patch-shaped designs where single and multiple split rings are engraved to identify the effect in return loss and radiation effect. These antennas are excited with the lumped plot element method with a rectangular port, as shown in Figure 1a. The FR4 sheet used to create the antenna is double-sided. The top and bottom layers are copper materials with perfect electric conductor properties. HFSS antenna modeling software was used to build the proposed structure. A squared patch is created by deducting the split ring from the required area. The top and bottom layers are ideal electric conductors with boundary conditions. Air is considered material when positioning the radiation box with the antenna. For the outer radiation box, we specify the radiation border condition. When the ground is used as a reference plane, the port depicted in Figure 1 is excited as a lumped port. The frequency sweep on the exciting port is what runs the simulation. Post-processing techniques are applied to the HFSS simulation of the antenna to extract the reflectance loss, gain, and directivity values.

Table 2. Different antenna designs and their descriptions.

Sr. No	Design	Description
1		Circular patch with stub in the back layer
2		Circular patch with split-ring-engraved structure with stub in the back layer
3		Circular patch with dual-split-ring resonator with a normal rectangular patch on the backside
4		Circular patch with dual-split-ring resonator with stub structure patch on the back side
5		Circular patch with 90-degree-rotated dual-split-ring resonator with the normal rectangular patch on the backside

3. Results and Discussion

The proposed antenna structure is numerically investigated and measured over the 1 to 25 GHz frequency range. We analyzed the different terms of the antenna, such as return loss parameters, gain, directivity, radiation pattern, efficiency, and electric field distribution. Figure 2 shows the representation of the co- and cross-polarization plot to the axis notation for the design 4 antenna with both port excitation conditions. Figure 3 shows the return loss values of the different S parameters (S_{11}, S_{12}, S_{21}, and S_{22}) (in dB) for design 1 and design 2. The S parameters are referred to as the power transfer ratio from one port to another. For example, the S parameters' term S_{ij} (i, j = 1, 2) refers to the ratio of the power transferred from jth port to ith port. Figure 3a shows the return loss values for design 1. The seven operating bands of the operation are observed in a simulation study. The top band of the operation is 3.9 GHz for the frequency range of 3.38 to 6.47 GHz. The return loss observed in this band is −13.81 dB. The minimum return loss of −18.41 dB in this antenna was observed between the 8.55 to 10.07 GHz frequency range. Figure 3b shows the return loss values for design 2. The five operating bands of 1.47 GHz, 0.66 GHz, 0.7 GHz, 1.06 GHz, and 5.38 GHz are observed in a simulation study. The maximum band of the operation is 5.38 GHz for a frequency range of 19.62 to 25 GHz. The return loss observed in this band is −25.91 dB.

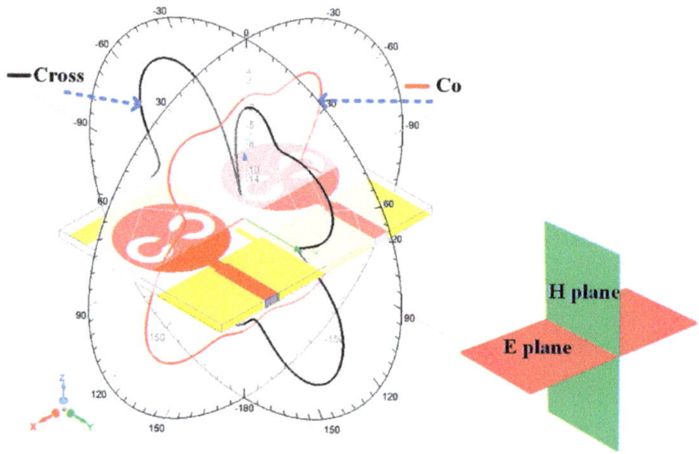

Figure 2. Representation of co−and cross−polarization plane to its axis system for design 4 antenna system with both port excitation conditions with E and H plane notation.

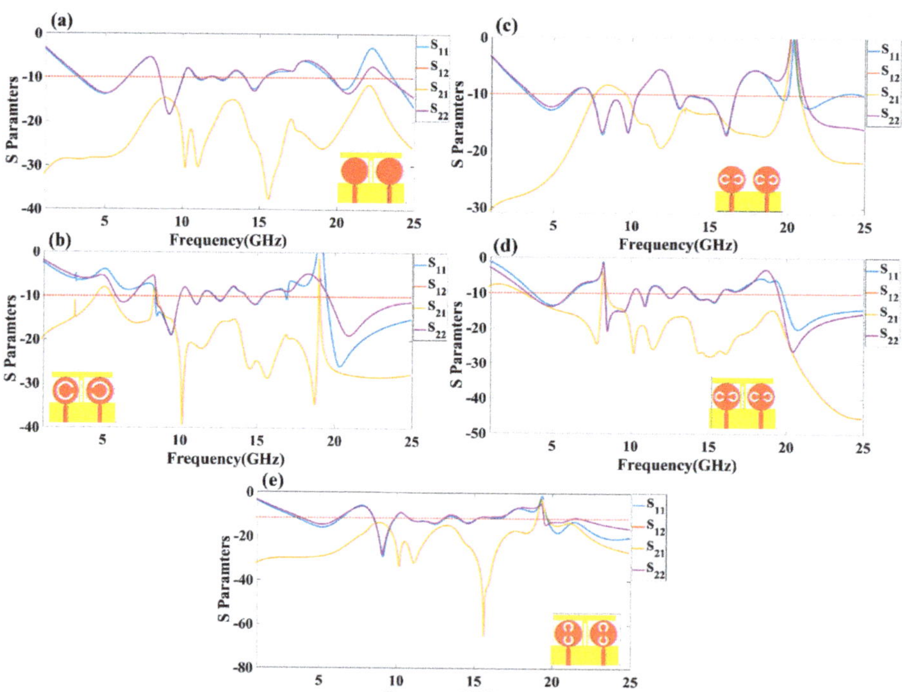

Figure 3. Calculated reflectance loss factors (S_{11}, S_{12}, S_{21}, and S_{22}) (in dB) for (**a**) design 1, (**b**) design 2, (**c**) design 3, (**d**) design 4, and (**e**) design 5 of the proposed structure of the antenna.

Figure 3c–e shows the return loss variation observed in design 3, design 4, and design 5. Figure 3a shows the return loss values for design 3. The six operating bands of 2.54 GHz, 0.29 GHz, 0.73 GHz, 3.94 GHz, 2.18 GHz, and 0.4 GHz are observed in a simulation study. The maximum band of the operation is 3.94 GHz for a frequency range of 12.75 to 16.69 GHz. The return loss observed in this band is −17.13 dB. Figure 3d shows the return loss values for design 4. The simulation study observed six operating bands of 2.46 GHz, 1.57 GHz, 0.51 GHz, 0.68 GHz, 1.73 GHz, and 5.18 GHz. The maximum band of the operation is 5.18 GHz for a frequency range of 19.82 to 25 GHz. The return loss observed in this band is −26.29 dB. Figure 3e shows the return loss values for design 5. The four operating bands of 3.42 GHz, 1.66 GHz, 6.55 GHz, and 5.28 GHz from the simulation study are observed. The maximum band of the operation is 5.28 GHz for a frequency range of 19.22 to 25 GHz. The return loss observed in this band is −20.84 dB. The overall minimum return loss of −29.33 dB is observed in the second band of 8.45 to 10.11 GHz frequency band. In Figure 2 of the S parameters' values of the antenna, the response of the S_{12} is overlapped with the parameter S_{21}. The reason for overlapping both responses is the ideally mirrored image on both sides of the antenna over the center axis.

The detailed comparative values in the operating frequency, a band of operation return loss, and the number of operating bands in each antenna are shown in Table 3. All the values for calculating the band and operating frequency are considered, where all the S parameters are <−10 dB. It is observed from all the designs that the engraved shape of the split-ring resonator and stub generate different operating bands with different return loss values. This result allows choosing the MIMO shape variant for the specific antenna operating band. The operating band is changing because the electric field concentrates on the different shapes and edges.

Table 3. The comparative study of achieved simulation results of frequency band and minimum return loss for the proposed antennas.

Design	f_{min}(GHz)	f_{max}(GHz)	Δf(GHz)	Minimum Return Loss (dB)
1	3.38	6.47	3.09	−13.81
	8.55	10.07	1.52	−18.41
	11.02	11.66	0.64	−10.58
	12.28	12.9	0.62	−10.67
	14.11	15.38	1.27	−13.03
	19.76	21.22	1.46	−13.45
	23.7	25	1.3	−16.56
2	8.57	10.04	1.47	−18.96
	10.76	11.42	0.66	−11.97
	12.24	12.94	0.7	−11.16
	14.32	15.38	1.06	−11.91
	19.62	25	5.38	−25.91
3	3.58	6.12	2.54	−12.82
	7.46	7.75	0.29	−12.77
	9.7	10.43	0.73	−16.8
	12.75	16.69	3.94	−17.13
	21.12	23.3	2.18	−15.43
	24.6	25	0.4	−16.04
4	3.75	6.21	2.46	−13.8
	8.56	10.13	1.57	−20.38
	10.65	11.16	0.51	−13.76
	12.26	12.94	0.68	−11.36
	14.06	15.79	1.73	−12.53
	19.82	25	5.18	−26.29
5	3.25	6.67	3.42	−15.97
	8.45	10.11	1.66	−29.33
	10.51	17.06	6.55	−14.29
	19.72	25	5.28	−20.84

Figure 4 shows the two-dimensional and three-dimensional radiation patterns for the design 1 antenna. Figure 4a,c,e shows the variation in three-dimensional radiation pattern for the different port excitation conditions of (P_1 = 0 and P_1 = 1), (P_1 = 1 and P_1 = 0), and (P_1 = 1 and P_1 = 1), respectively. Similarly, Figure 5a–c shows the variation in two-dimensional radiation pattern for the different port excitation conditions of (P_1 = 0 and P_1 = 1), (P_1 = 1 and P_1 = 0), and (P_1 = 1 and P_1 = 1), respectively. Similarly, Figure 4d–o shows the three-dimensional and two-dimensional radiation patterns for design 2, design 3, design 4, and design 5 of antennas. Notably, variation in the antenna radiation pattern is observed for the different port conditions along with the peak gain achieved. The >4 dBi gain is observed in all the antenna designs with different port conditions.

We calculated the half-power beamwidth of the different antenna for both port excitation conditions, and its comparative analysis is shown in Table 4. The beam width variation is observed from a narrow beam to a wide beam (HPBW), ranging from 2° to 75°. The comparative analysis is presented in Table 4 in terms of design, polarization conditions, and HPBM range. HPBM is calculated from the normalized bandwidth of the overall pattern to >−3 dB values, which are referred to as half-power range. It is observed from Figure 5a,b that the left-hand plane offers a wide beam width compared with the right-hand plane. As shown in Figure 5c and Table 4 (Design 1), the maximum and minimum values of HPBW are 44° and 7° in co-polarization conditions. In cross-polarization conditions, the maximum 36° of the HPBW is also observed in −63° to −27° of the radiation plane. In the radiation pattern of design 2 shown in Figure 5f, we observed a maximum HPBW of 42°. The addition of the split ring in the top MIMO patch increases the beam width of the directivity. Similarly, the wide beam width of 75° in the double split-ring resonator structure, as shown in Figure 5i for design 3 of the antenna structure, is observed. The effect of adding the stub in the bottom layer increases the radiation beam on the overall radiation patterns. The stub in the middle converts the majority of single-beam radiation to a multibeam radiation pattern. The effect of this multibeam radiation is observed in Figure 5l for design 4 of the antenna structure. The maximum and minimum beamwidths in the radiation pattern are observed as 62° and 5° in the co-polarization pattern. Similarly, the maximum and minimum beamwidths in the radiation pattern are observed as 37° and 2° in the co-polarization pattern. The double split-ring orientation change generates the maximum and minimum HPBWs of 46° and 3° for the co-polarization structure for design 5, as shown in Figure 5o. In the cross-polarization radiation pattern, the maximum and minimum HPBWs are 38° and 13°. The effect of the split ring is mainly observed for shaping the multibeam antenna radiation structure by applying specific ring orientation. The phase difference between co- and cross-polarization antennas is 0° when peak gain values are high. These conditions are observed for all port excitation conditions. It will make the antenna behavior be in linear polarization mode. This phase difference can be observed in Figure 4c,f,i,l,o.

We also calculated the variation in directivity for the entire frequency spectrum band, and the results are shown in Figure 6. Figure 6 shows the change in normalized directivity (dB) for the angle values of −180° to 180° and phi of radiation pattern 0 degrees for the entire frequency spectrum of 1 to 25 GHz. We can also identify the values of the >−3 dB operating band for the specific radiation rather subgraph and its suitable frequency bands. Figure 7 shows the radiation efficiency of the different antenna designs over the entire frequency band of the operation. The variation in frequency excitation from 1 to 25 GHz values of this process gives us the efficiency values of the spectrum bands. The maximum radiation efficiency peak shows near the 8 GHz and 18 GHz frequency points. Comparing the return loss values and gain plot, these two points do not show the practical return loss values. In other bands of the operation, as shown in Figure 3, the average radiation efficiency is observed at > 40% for most bands. Therefore, we can choose the effective radiation efficiency, return loss values, gain, and directivity values to identify the effective operation of the MIMO antenna. The radiation efficiency allows for identifying the broad operating frequency band where the antenna radiated to its higher amplitude of electromagnetic radiation. The combined results of the return loss and efficiency allow for choosing the type of design, operating frequency spectrum, and radiation bandwidth.

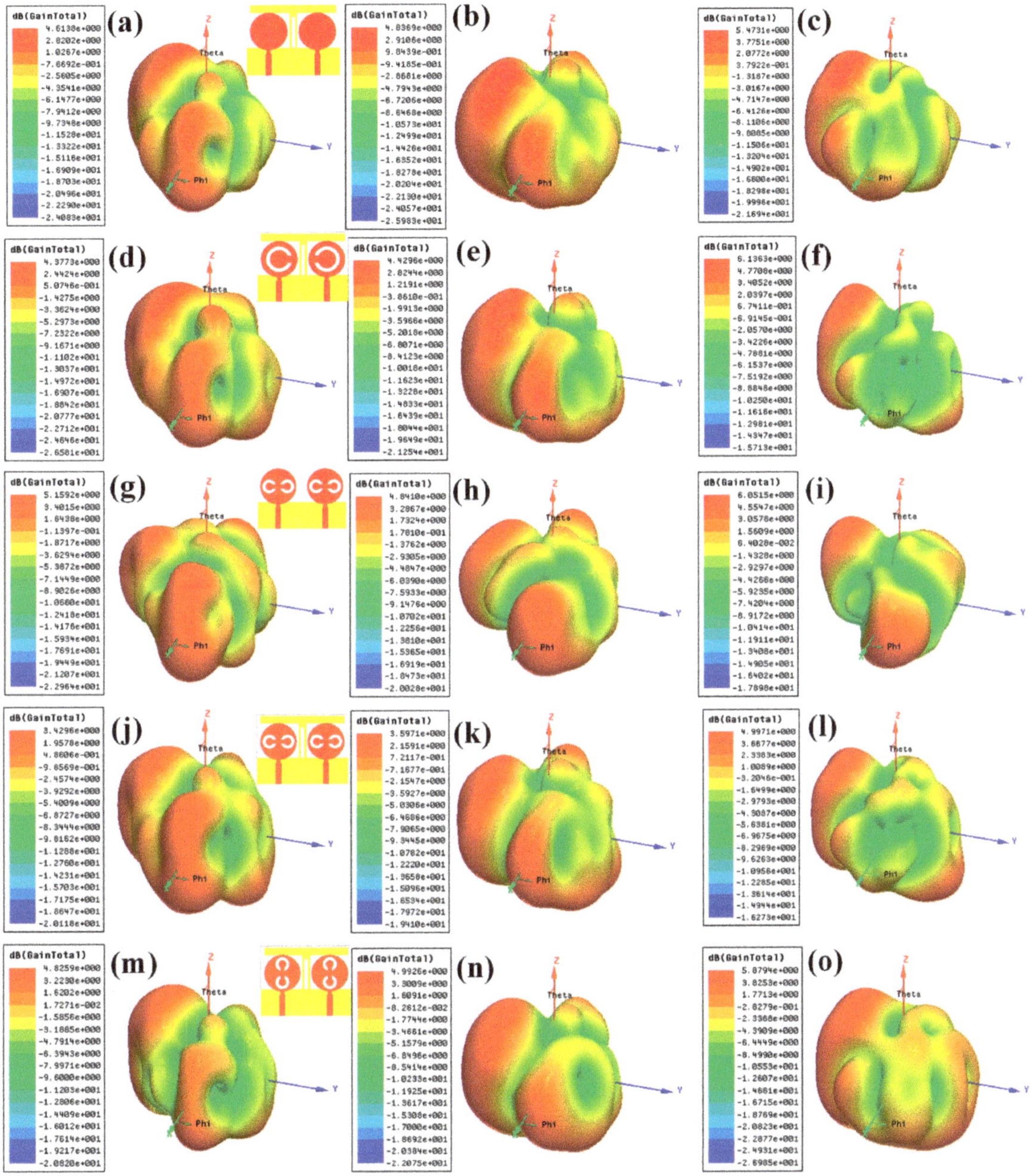

Figure 4. Variation in the radiation in 3D polar plots for the different port excitation conditions of (**a**–**c**) design 1, (**d**–**f**) design 2, (**g**–**i**) design 3, (**j**–**l**) design 4, and (**m**–**o**) design 5 antenna. The different port excitation conditions are (**a**,**d**,**g**,**j**,**m**) $P_1 = 0$ and $P_2 = 1$, (**b**,**e**,**h**,**k**,**n**) $P_1 = 1$ and $P_2 = 0$, and (**c**,**f**,**i**,**l**,**o**) $P_1 = 1$ and $P_2 = 1$.

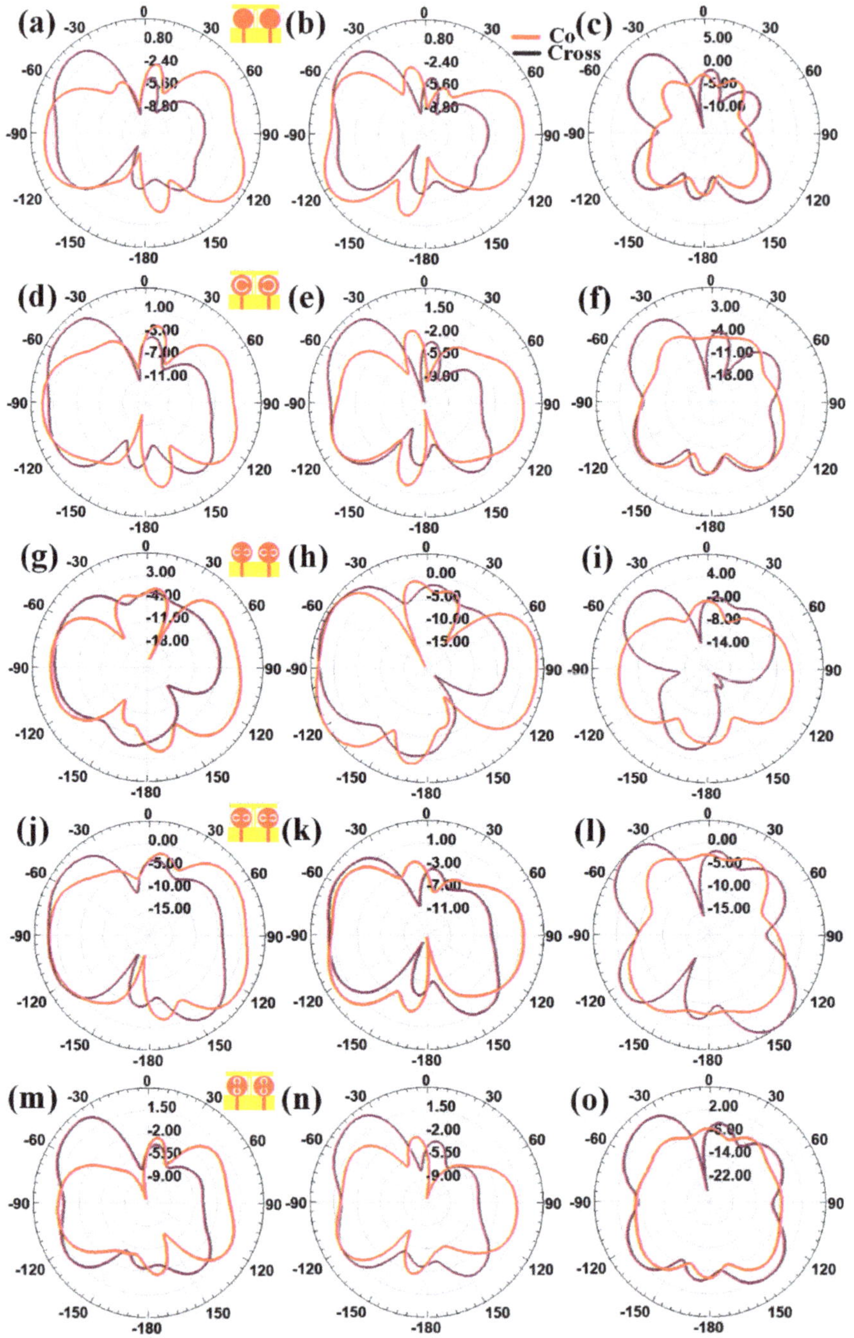

Figure 5. Variation in the radiation in 2D polar plots for the different port excitation conditions of (**a–c**) design 1, (**d–f**) design 2, (**g–i**) design 3, (**j–l**) design 4, and (**m–o**) design 5 antenna. The different port excitation conditions are (**a,d,g,j,m**) $P_1 = 0$ and $P_2 = 1$, (**b,e,h,k,n**) $P_1 = 1$ and $P_2 = 0$, and (**c,f,i,l,o**) $P_1 = 1$ and $P_2 = 1$.

Table 4. The comparative analysis of the calculated half-power beamwidth (HPBW) for the antennas at both port excitation conditions $P_1 = 1$ and $P_2 = 1$.

Design	Polarization	HPBW Range (Degree)	HPBW (Degree)
1	Co	−180 to −173	7
		−154 to −110	44
		−45 to −26	19
		−6 to 7	13
		25 to 46	21
		110 to 153	43
		170 to 180	10
	Cross	−63 to −27	36
		129 to 146	17
2	Co	−148 to −106	42
		105 to 149	42
	Cross	−63 to −26	37
3	Co	−134 to −59	75
		62 to 134	72
	Cross	−67 to −30	37
4	Co	−180 to −175	5
		−164 to −104	60
		−50 to −10	40
		16 t0 50	24
		103 to 165	62
		175 to 180	5
	Cross	−126 to −124	2
		−65 to −29	36
		121 to 158	37
5	Co	−180 to −177	3
		−152 to −106	46
		32 to 43	9
		110 to 152	42
		177 to 180	3
	Cross	−64 to −26	38
		133 to 146	13

The peak gain achieved in all the antenna ranges for the 1 to 25 GHz of the frequency spectrum is presented in Figure 8. It is observed from the combined return loss and gains spectrum that the peak gain is not the only factor in overall antenna operation that can justify. It is essential to look first for the fair return loss values, antenna efficiency, and gain value to judge the overall performance of the proposed antennas. In some bands, the return loss is >−10 dB, but gain values seem higher. The peak near to 20 GHz frequency values are >−10 dB. This combination is not helpful for effective antenna radiation. This type of condition leads toward the oscillating behavior of the antenna instead of radiating behavior. Figure 9 shows the different antenna designs' electric field (V/m) distribution on radiation planes. It is observed that there are different field distributions over different shapes and engraved structures. Variation in the electric field distribution is changing per the split ring resonator based engraved design and substrate stub design. The electric field distribution over the different shape structures creates different resonator points of frequency. It will affect variation in reflectance parameters and gain values. Figures 10 and 11 show the variation in the reflectance loss between simulated and measured results of the proposed antenna structures. This MIMO antenna design has a very small resonating point variation. The feed points are the primary ones responsible for the significant variation in the measured and simulated results for specific operating bands. The

SMA connector adds a significant amount of loss to the measurements. The resonator's bandwidth, however, is roughly the same in both cases. Measured and simulated results for these MIMO antennas show significant variation, which can be attributed in part to the effect of the soldering materials. Either result's alteration may also be caused by the connector employed during the measurement set. The impedance of the connector or the antenna may cause this difference. The MIMO antenna parameters such as ECC (Envelope Correlation Coefficient), CCL (Channel Capacity Loss), MEG (Mean Effective Gain), DG (Directivity Gain), and TARC (Total Active Reflection Coefficient) [5,32–34] were investigated for the proposed structure 4 antenna. The achieved diversified MIMO parameter results of the proposed design 4 antenna show the values of ECC < 0.01, CCL < 0.5 bps/Hz/s, DG $\approx$ 10 dB, TARC< -10.0 dB, and MEG < 0.5 dB). It helps overcome the obstacles inherent in short-distance communication, such as fading signals, increased interference, and The measured antenna efficiency for the proposed antenna structure is shown in Figure 12. In the simulation, efficiency is measured by defaulted simulation setup and parameter extraction provided by the HFSS simulation package. The ratio of the received power and transmitted power extracts the efficiency of the measured antenna. multipath propagation. These MIMO antenna parameters for the design 4 antenna structure are shown in Figure 13. The comparative analysis of the proposed structure with the previously published results is shown in Table 5. The comparison of the proposed structure is derived in terms of the isolation, ECC, antenna structure area, antenna elements, and design complexity. Figure 14 shows the variation in cross-port reflectance loss values for the different antenna designs. The comparative analysis of the S12 values is majority <-10 dB in the whole spectrum for most antenna designs. This response suggests that all the antenna generates the minimum cross-port interaction as the values are <-10 dB.

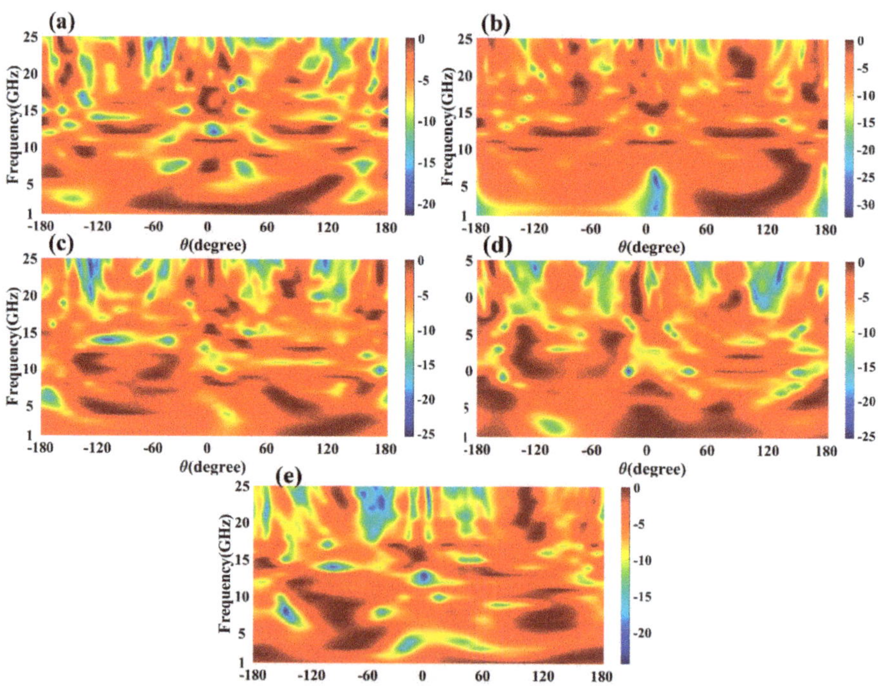

Figure 6. Calculated variation in normalized gain (dBi) for the different polar angles and frequency spectrums. Variation in gain for the different (**a**) design 1, (**b**) design 2, (**c**) design 3, (**d**) design 4, and (**e**) design 5 antennas.

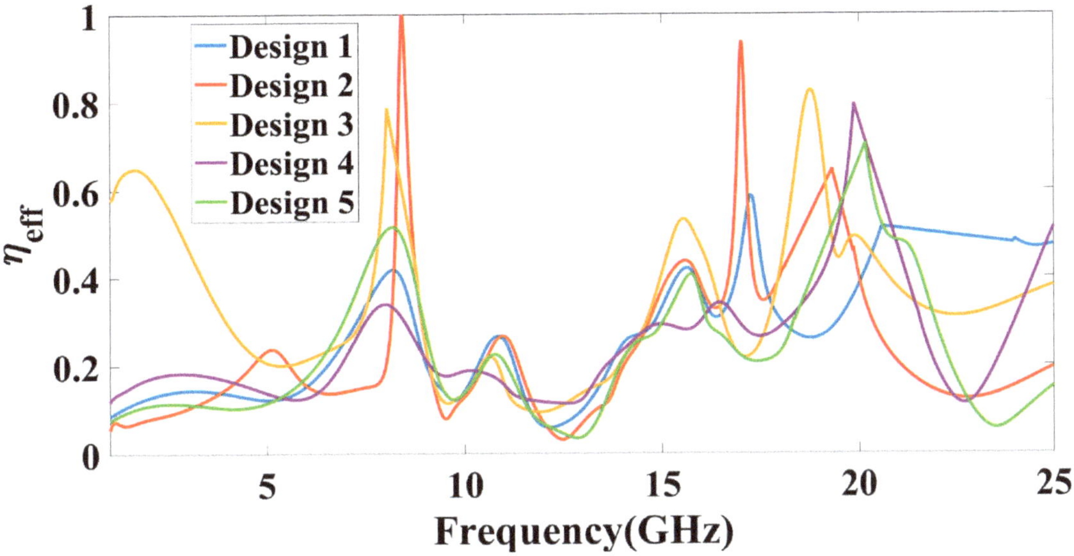

Figure 7. Variation in antenna radiation efficiency for different designs of antenna.

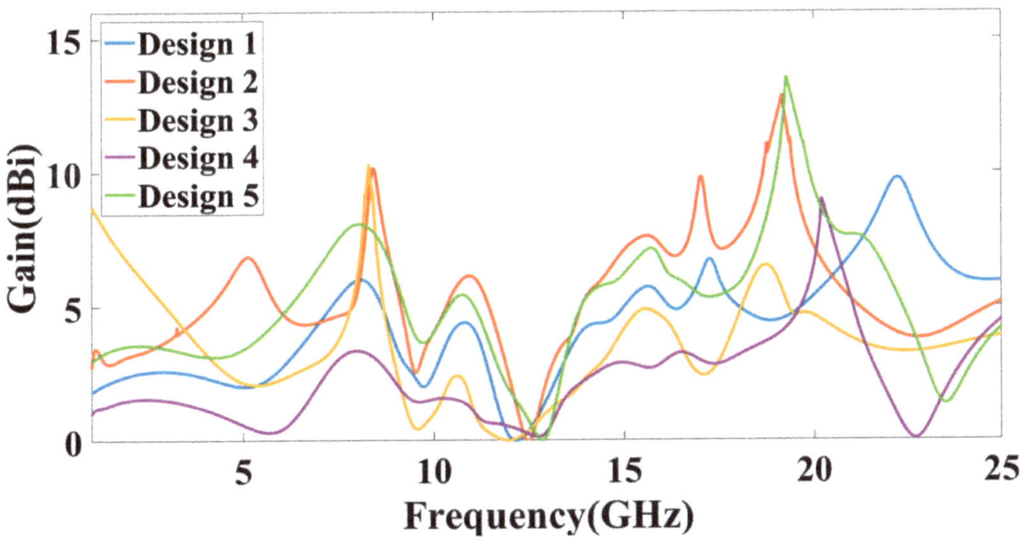

Figure 8. Variation in peak gain (dBi) for the different antenna designs.

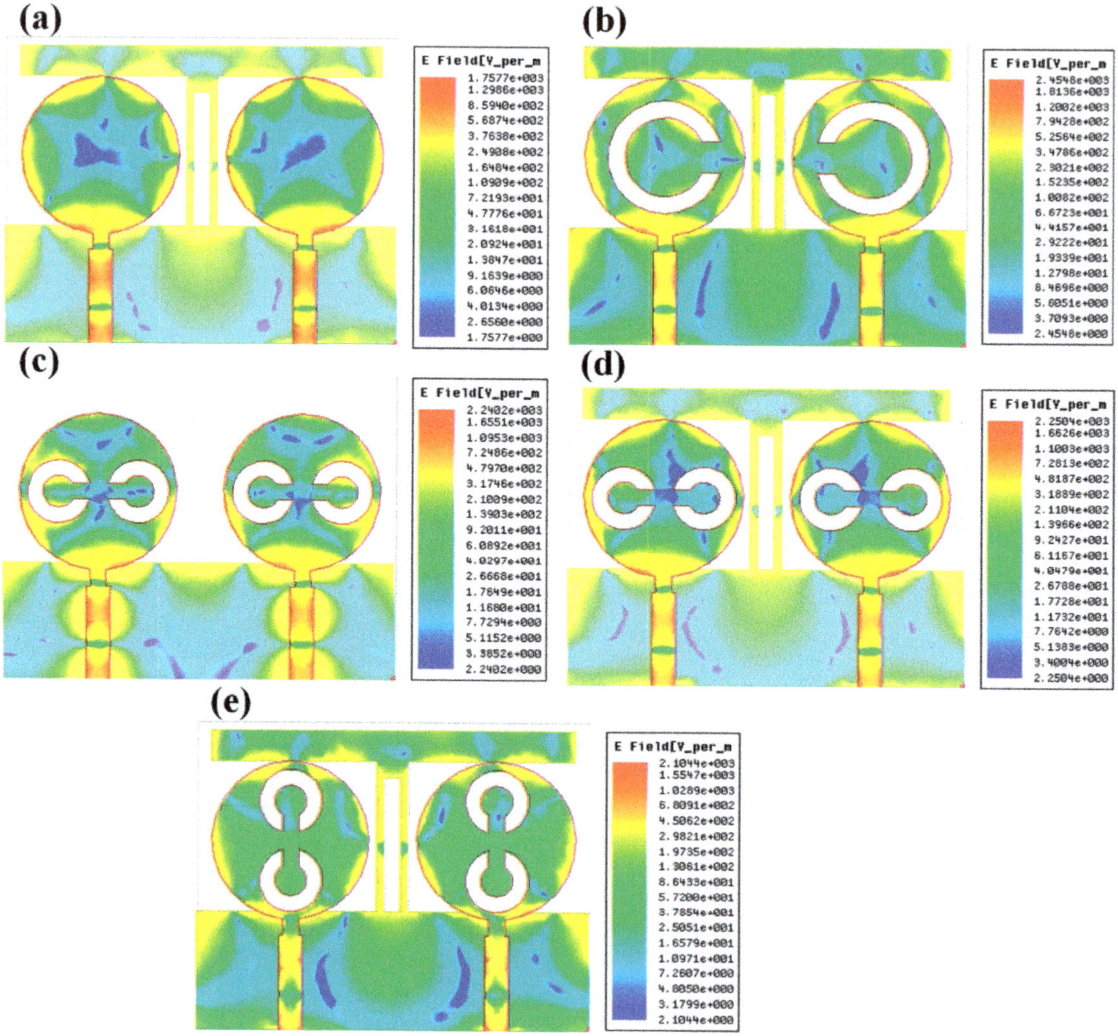

Figure 9. Normalized electric field distribution for the different shapes of the structures. The normalized electric field intensity for the different antenna designs (**a**) Design 1, (**b**) Design 2, (**c**) Design 3, (**d**) Design 4, and (**e**) Design 5 for the 10 GHz frequency value.

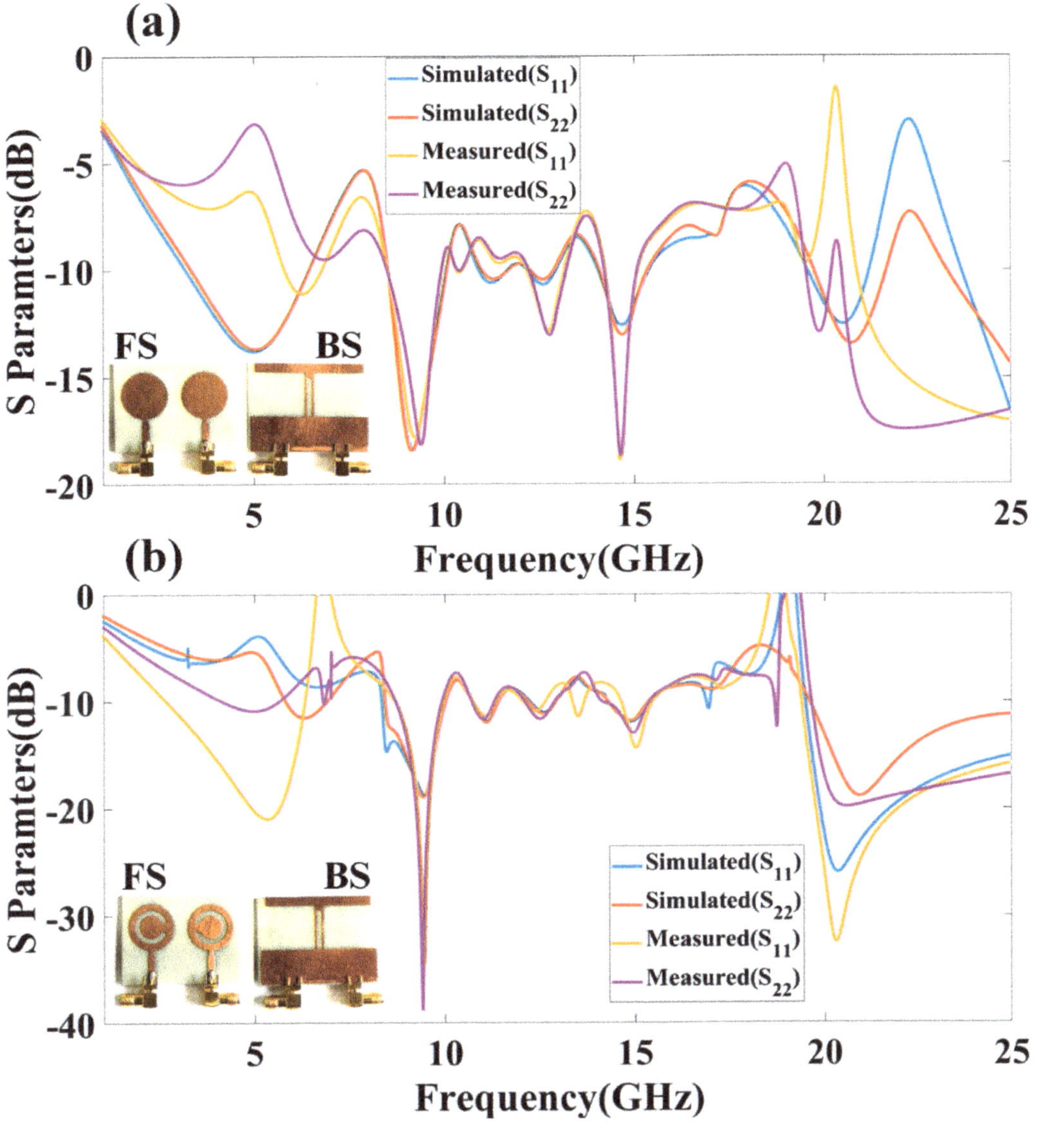

Figure 10. Comparative plot of simulated and measured reflectance loss factors (S_{11}, S_{22}) (in dB) for (**a**) design 1 and (**b**) design 2 of the proposed structure of the antenna.

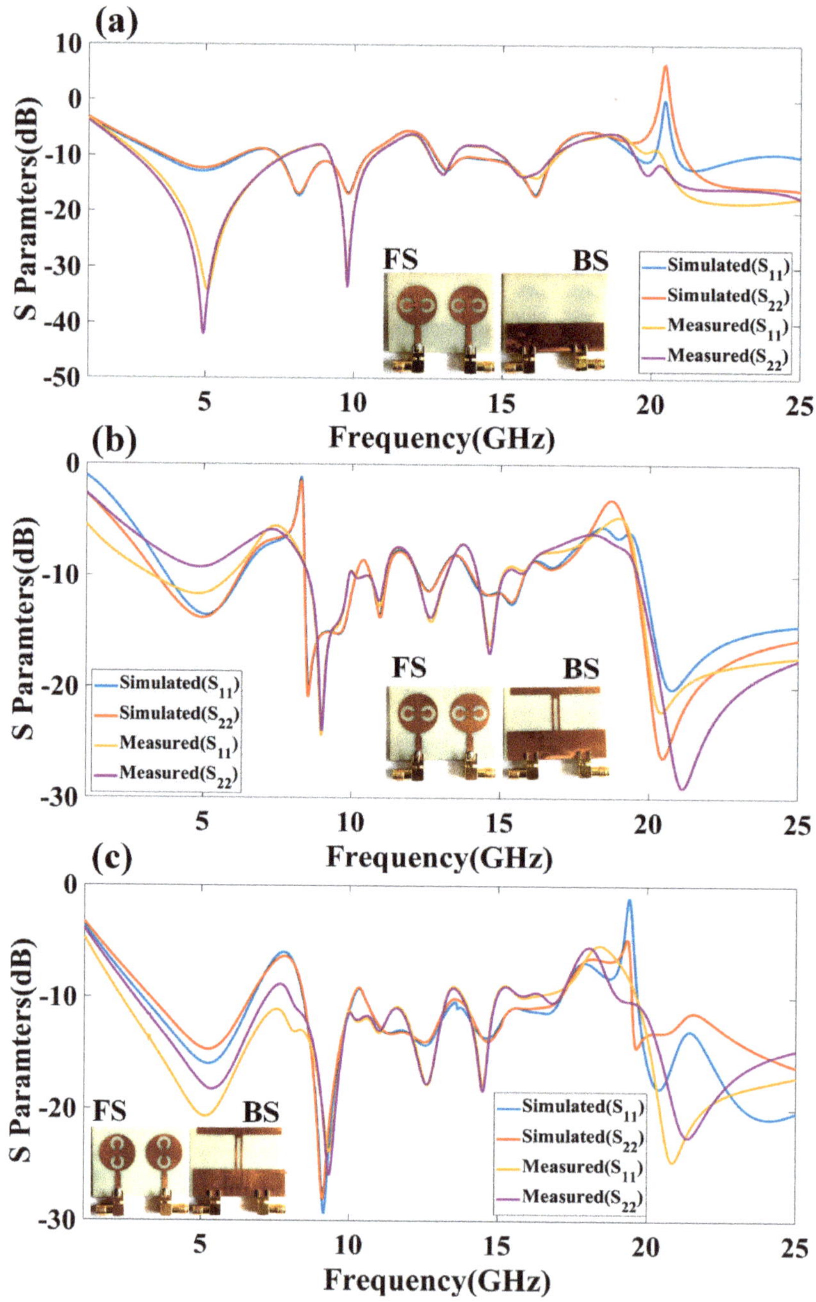

Figure 11. Comparative plot of simulation and measured reflectance loss factors (S_{11}, S_{22}) (in dB) for (**a**) design 3, (**b**) design 4, and (**c**) design 5 of the proposed structure of the antenna.

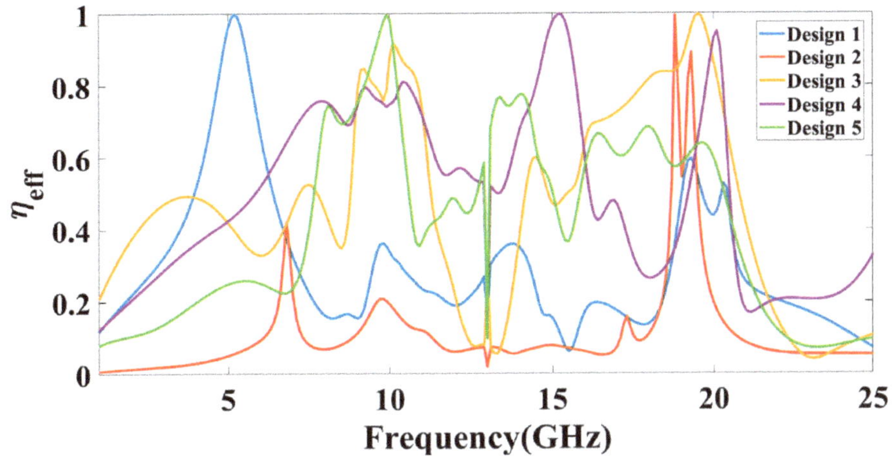

Figure 12. Variation in antenna radiation efficiency (measured) for different antenna designs.

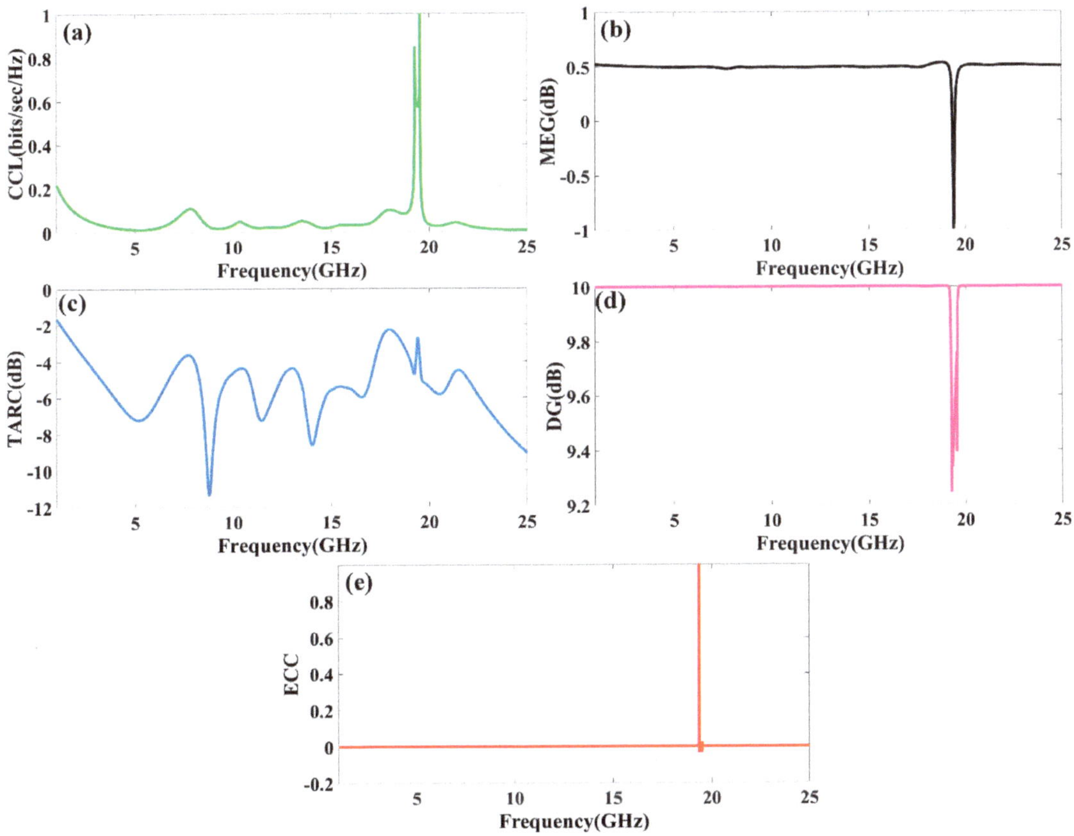

Figure 13. Variation in different MIMO antenna parameters (**a**) CCL, (**b**) MEG, (**c**) TARC, (**d**) DG, and (**e**) ECC for the design 4 antenna structure.

Table 5. Comparative analysis of the proposed MIMO antenna with previously published results in terms of different antenna parameters.

No of Elements	Isolation (in dB)	Design Complexity	Antenna Area (mm^2)	ECC	Value of Peak Gain (dBi)	Reference
2	25	No	1950	<0.1	>10	This structure
4	21	No	2460	<0.25	6.5	[29]
2	20	No	4371	-	3.5	[35]
4	10	No	2025	<0.25	2.75	[19]
4	17	Yes	1600	0.06	2.9	[36]
4	15	No	1600	0.4	3.5	[37]
2	22	No	640	-	1	[21]
4	10	Yes	5184	<0.014	3.1	[38]
2	21	No	680	-	4	[22]
4	13	No	5624	<0.04	6.2	[39]
2	-	Yes	2880	<0.02	3	[40]
4	13	Yes	5625	<0.04	6.2	[41]
4	15	Yes	16,800	<0.10	5.5	[42]
2	15	Yes	13,125	<0.15	5.1	[43]
4	11	No	1600	<0.1	4	[18]
4	15.4	Yes	676	<0.01	1.41	[20]
4	22	Yes	1750	0.003	-	[44]

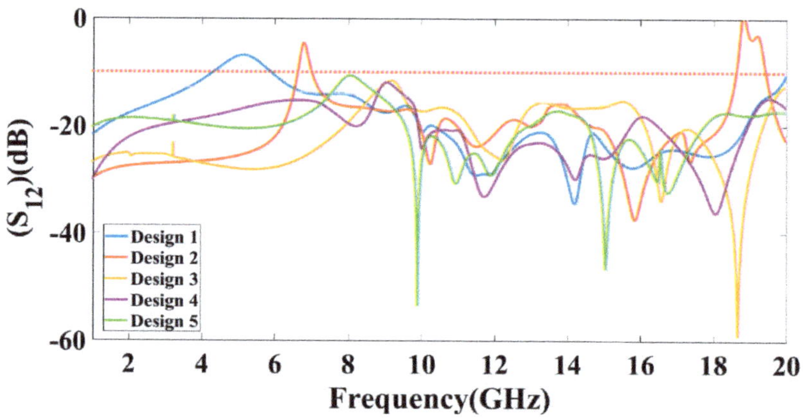

Figure 14. Comparative plot of measured cross-port reflection values (S_{12}) for the different antenna designs.

4. Conclusions

We presented numerical and measured results of the split ring resonator loaded antenna for the 1 to 25 GHz frequency range. We showed the different antenna parameter results for the five antenna types, which contain the simple circular patch shape, single split-ring, and double split-ring resonator structures. The antenna is analyzed in terms of the different reflectance parameters, gain, directivity, efficiency, peak gain, and electric field distribution. The proposed antenna offers a maximum bandwidth of 3.09 GHz with −13.81 dB return loss in a normal circular patch antenna. In the case of the single split-ring resonator, we observed the 5.38 GHz bandwidth with −25.91 dB of return loss. The double split-ring resonator structure offers the maximum bandwidth of 5.28 GHz with −20.84 dB of the return loss. We observed the bandwidth, return loss, gain, and bandwidth variation for all antenna structures. The proposed antenna's overall results can apply to various Wi-Fi, WiMAX, and 5G communication applications.

Author Contributions: Conceptualization, A.A. and V.S.; methodology A.A. and V.S.; software, validation, formal analysis, investigation, and writing—original draft preparation K.A., F.A. (Fayadh Alenezi), M.A. and F.A. (Farman Ali); writing—review and editing., K.A., F.A. (Fayadh Alenezi), M.A. and F.A. (Farman Ali); supervision, A.A. and V.S.; project administration, A.A.; funding acquisition, A.A. All authors have read and agreed to the published version of the manuscript.

Funding: This work was funded by the Deanship of Scientific Research at Jouf University under grant number (DSR2022-RG-0111).

Data Availability Statement: Data are available based upon reasonable request from corresponding author.

Acknowledgments: The authors would like to thank the Deanship of Scientific Research at Jouf University for the support.

Conflicts of Interest: Authors declare no conflict of interest.

References

1. Fathima, N.; Nayana, K.S.; Ali, T.; Biradar, R.C. A miniaturized slotted ground fractal Koch multiband antenna for wireless applications. In Proceedings of the RTEICT 2017-2nd IEEE International Conference on Recent Trends in Electronics, Information and Communication Technology, Bangalore, India, 19–20 May 2017; pp. 251–255. [CrossRef]
2. Balani, W.; Sarvagya, M.; Ali, T.; Pai, M.; Anguera, J.; Andujar, A.; Das, S. Design Techniques of Super-Wideband Antenna–Existing and Future Prospective. *IEEE Access* **2019**, *7*, 141241–141257. [CrossRef]
3. JPark, J.-D.; Rahman, M.; Chen, H.N. Isolation Enhancement of Wide-Band MIMO Array Antennas Utilizing Resistive Loading. *IEEE Access* **2019**, *7*, 81020–81026. [CrossRef]
4. Malekpour, N.; Honarvar, M.A. Design of high-isolation compact MIMO antenna for UWB application. *Prog. Electromagn. Res. C* **2016**, *62*, 119–129. [CrossRef]
5. Alharbi, A.G.; Sorathiya, V. Ultra-Wideband Graphene-Based Micro-Sized Circular Patch-Shaped Yagi-like MIMO Antenna for Terahertz Wireless Communication. *Electronics* **2022**, *11*, 1305. [CrossRef]
6. Ren, J.; Hu, W.; Yin, Y.; Fan, R. Compact Printed MIMO Antenna for UWB Applications. *IEEE Antennas Wirel. Propag. Lett.* **2014**, *13*, 1517–1520. [CrossRef]
7. Dave, K.; Sorathiya, V.; Lavadiya, S.P.; Patel, S.K.; Dhankecha, U.; Swain, D.; Faragallah, O.S.; Eid, M.M.A.; Rashed, A.N.Z. Graphene-based double-loaded complementary split ring resonator (CSRR) slotted MIMO patch antenna for spectroscopy and imaging THz applications. *Appl. Phys. A* **2022**, *128*, 656. [CrossRef]
8. Jensen, M.; Wallace, J. A Review of Antennas and Propagation for MIMO Wireless Communications. *IEEE Trans. Antennas Propag.* **2004**, *52*, 2810–2824. [CrossRef]
9. Saadh, A.M.; Ashwath, K.; Ramaswamy, P.; Ali, T.; Anguera, J. A uniquely shaped MIMO antenna on FR4 material to enhance isolation and bandwidth for wireless applications. *AEU-Int. J. Electron. Commun.* **2020**, *123*, 153316. [CrossRef]
10. Li, Y.; Li, W.; Yu, W. A multi-band/UWB MIMO/diversity antenna with an enhanced isolation using radial stub loaded resonator. *Appl. Comput. Electromagn. Soc. J.* **2013**, *28*, 8–20.
11. Zhao, Y.; Li, Y.; Shi, W.; Yu, W. Mutual coupling reduction between patch antenna and microstrip transmission line by using defected isolation wall. *Appl. Comput. Electromagn. Soc. J.* **2019**, *34*, 1.
12. Aw, M.S.; Ashwath, K.R.; Ali, T. A compact two element MIMO antenna with improved isolation for wireless applications. *J. Instrum.* **2019**, *14*, P06014. [CrossRef]
13. BharathiDevi, B.; Kumar, J. Small frequency range discrete bandwidth tunable multiband MIMO antenna for radio/LTE/ISM-2.4 GHz band applications. *AEU-Int. J. Electron. Commun.* **2021**, *144*, 154060. [CrossRef]
14. Lavadiya, S.P.; Sorathiya, V.; Kanzariya, S.; Chavda, B.; Naweed, A.; Faragallah, O.S.; Eid, M.M.A.; Rashed, A.N.Z. Low profile multiband microstrip patch antenna with frequency reconfigurable feature using PIN diode for S, C, X, and Ku band applications. *Int. J. Commun. Syst.* **2022**, *35*, 9. [CrossRef]
15. Hasan, M.; Islam, M.T.; Samsuzzaman; Baharuddin, M.H.; Soliman, M.S.; Alzamil, A.; Abu Sulayman, I.I.M.; Islam, S. Gain and isolation enhancement of a wideband MIMO antenna using metasurface for 5G sub-6 GHz communication systems. *Sci. Rep.* **2022**, *12*, 9433. [CrossRef]
16. Megahed, A.A.; Abdelazim, M.; Abdelhay, E.H.; Soliman, H.Y.M. Sub-6 GHz Highly Isolated Wideband MIMO Antenna Arrays. *IEEE Access* **2022**, *10*, 19875–19889. [CrossRef]
17. Jabeen, S.; Khan, Q.U. An integrated MIMO antenna design for Sub-6 GHz & millimeter-wave applications with high isolation. *AEU-Int. J. Electron. Commun.* **2022**, *153*, 154247. [CrossRef]
18. Sarkar, D.; Srivastava, K.V. A compact four-element MIMO/diversity antenna with enhanced bandwidth. *IEEE Antennas Wirel. Propag. Lett.* **2017**, *16*, 2469–2472. [CrossRef]
19. Anitha, R.; Vinesh, P.V.; Prakash, K.C.; Mohanan, P.; Vasudevan, K. A Compact Quad Element Slotted Ground Wideband Antenna for MIMO Applications. *IEEE Trans. Antennas Propag.* **2016**, *10*, 4550–4553. [CrossRef]

20. Pandit, S.; Mohan, A.; Ray, P. A compact four-element MIMO antenna for WLAN applications. *Microw. Opt. Technol. Lett.* **2018**, *60*, 289–295. [CrossRef]
21. Azarm, B.; Nourinia, J.; Ghobadi, C.; Majidzadeh, M. Highly isolated dual band stop two-element UWB MIMO antenna topology for wireless communication applications. *J. Instrum.* **2019**, *14*, P10036. [CrossRef]
22. Hatami, N.; Nourinia, J.; Ghoabdi, C.; Majidzadeh, M.; Azarm, B. High inter-element isolation and WLAN filtering mechanism: A compact MIMO antenna scheme. *AEU-Int. J. Electron. Commun.* **2019**, *109*, 43–54. [CrossRef]
23. Li, H.; Shi, D.; Wang, W.; Liao, D.; Gadekallu, T.R.; Yu, K. Secure routing for LEO satellite network survivability. *Comput. Netw.* **2022**, *211*, 109011. [CrossRef]
24. Wang, W.; Xu, H.; Alazab, M.; Gadekallu, T.R.; Han, Z.; Su, C. Blockchain-Based Reliable and Efficient Certificateless Signature for IIoT Devices. *IEEE Trans. Ind. Inform.* **2021**, *18*, 7059–7067. [CrossRef]
25. Ding, F.; Zhu, G.; Alazab, M.; Li, X.; Yu, K. Deep-Learning-Empowered Digital Forensics for Edge Consumer Electronics in 5G HetNets. *IEEE Consum. Electron. Mag.* **2020**, *11*, 42–50. [CrossRef]
26. Naqvi, S.I.; Hussain, N.; Iqbal, A.; Rahman, M.; Forsat, M.; Mirjavadi, S.S.; Amin, Y. Integrated LTE and Millimeter-Wave 5G MIMO Antenna System for 4G/5G Wireless Terminals. *Sensors* **2020**, *20*, 3926. [CrossRef] [PubMed]
27. Dong, Y.; Liu, P.; Yu, D.; Li, G.; Tao, F. Dual-Band Reconfigurable Terahertz Patch Antenna with Graphene-Stack-Based Backing Cavity. *IEEE Antennas Wirel. Propag. Lett.* **2016**, *15*, 1541–1544. [CrossRef]
28. Li, J.; Zhang, X.; Wang, Z.; Chen, X.; Chen, J.; Li, Y.; Zhang, A. Dual-Band Eight-Antenna Array Design for MIMO Applications in 5G Mobile Terminals. *IEEE Access* **2019**, *7*, 71636–71644. [CrossRef]
29. Liu, X.-L.; Wang, Z.-D.; Yin, Y.-Z.; Ren, J.; Wu, J.-J. A Compact Ultrawideband MIMO Antenna Using QSCA for High Isolation. *IEEE Antennas Wirel. Propag. Lett.* **2014**, *13*, 1497–1500. [CrossRef]
30. Chae, S.H.; Oh, S.-K.; Park, S.-O. Analysis of Mutual Coupling, Correlations, and TARC in WiBro MIMO Array Antenna. *IEEE Antennas Wirel. Propag. Lett.* **2007**, *6*, 122–125. [CrossRef]
31. Zahoor, M.; Dou, Z.; Shah, S.; Khan, I.; Ayub, S.; Gadekallu, T.R. Pilot Decontamination Using Asynchronous Fractional Pilot Scheduling in Massive MIMO Systems. *Sensors* **2020**, *20*, 6213. [CrossRef]
32. Babu, K.V.; Anuradha, B. Design of inverted L-shape & ohm symbol inserted MIMO antenna to reduce the mutual coupling. *AEU-Int. J. Electron. Commun.* **2019**, *105*, 42–53. [CrossRef]
33. Babu, K.V.; Anuradha, B.; Das, S. Design & analysis of a dual-band MIMO antenna to reduce the mutual coupling. *J. Instrum.* **2019**, *14*, P09023. [CrossRef]
34. Krishna, C.M.; Das, S.; Lakrit, S.; Lavadiya, S.; Madhav, B.T.P.; Sorathiya, V. Design and analysis of a super wideband (0.09–30.14 THz) graphene based log periodic dipole array antenna for terahertz applications. *Optik* **2021**, *247*, 167991. [CrossRef]
35. Radhi, A.H.; Nilavalan, R.; Wang, Y.; Al-Raweshidy, H.S.; Eltokhy, A.A.; Ab Aziz, N. Mutual coupling reduction with a wideband planar decoupling structure for UWB–MIMO antennas. *Int. J. Microw. Wirel. Technol.* **2018**, *10*, 1143–1154. [CrossRef]
36. Rajkumar, S.; Amala, A.A.; Selvan, K. Isolation improvement of UWB MIMO antenna utilising molecule fractal structure. *Electron. Lett.* **2019**, *55*, 576–579. [CrossRef]
37. Khan, A.A.; Naqvi, S.A.; Khan, M.S.; Ijaz, B. Quad port miniaturized MIMO antenna for UWB 11 GHz and 13 GHz frequency bands. *AEU-Int. J. Electron. Commun.* **2021**, *131*, 153618. [CrossRef]
38. Chouhan, S.; Panda, D.K.; Kushwah, V.S. Modified circular common element four-port multiple-input-multiple-output antenna using diagonal parasitic element. *Int. J. RF Microw. Comput. Eng.* **2018**, *29*, e21527. [CrossRef]
39. Malviya, L.; Chouhan, S. Multi-cut four-port shared radiator with stepped ground and diversity effects for WLAN application. *Int. J. Microw. Wirel. Technol.* **2019**, *11*, 1044–1053. [CrossRef]
40. Jaglan, N.; Gupta, S.D.; Thakur, E.; Kumar, D.; Kanaujia, B.K.; Srivastava, S. Triple band notched mushroom and uniplanar EBG structures based UWB MIMO/Diversity antenna with enhanced wide band isolation. *AEU-Int. J. Electron. Commun.* **2018**, *90*, 36–44. [CrossRef]
41. Chouhan, S.; Malviya, L. Four-port shared rectangular radiator with defected ground for wireless application. *Int. J. Commun. Syst.* **2020**, *33*, e4356. [CrossRef]
42. MoradiKordalivand, A.; Rahman, T.A.; Khalily, M. Common Elements Wideband MIMO Antenna System for WiFi/LTE Access-Point Applications. *IEEE Antennas Wirel. Propag. Lett.* **2014**, *13*, 1601–1604. [CrossRef]
43. Moradikordalivand, A.; Leow, C.Y.; Rahman, T.A.; Ebrahimi, S.; Chua, T.H. Wideband MIMO antenna system with dual polarization for WiFi and LTE applications. *Int. J. Microw. Wirel. Technol.* **2015**, *8*, 643–650. [CrossRef]
44. Chithradevi, R.; Sreeja, B.S. A compat UWB MIMO antenna with high isolation and low correlation for wireless applications. In Proceedings of the 2017 IEEE International Conference on Antenna Innovations & Modern Technologies for Ground, Aircraft and Satellite Applications (iAIM), Bangalore, India, 24–26 November 2017; pp. 1–4. [CrossRef]

micromachines

Article

Investigations on Stub-Based UWB-MIMO Antennas to Enhance Isolation Using Characteristic Mode Analysis

Ankireddy Chandra Suresh [1], Thatiparthi Sreenivasulu Reddy [1], Boddapati Taraka Phani Madhav [2], Sudipta Das [3,*], Sunil Lavadiya [4], Abeer D. Algarni [5] and Walid El-Shafai [6,7]

1 Department of Electronics and Communication Engineering, Sri Venkateswara University College of Engineering, SV University, Tirupathi 517502, A.P, India
2 Antennas and Liquid Crystals Research Center, Department of Electronics and Communication Engineering, Koneru Lakshmaiah Education Foundation, Vaddeswaram 522303, India
3 Department of Electronics & Communication Engineering, IMPS College of Engineering and Technology, Malda 732103, India
4 Department of Information and Communication Technology, Marwadi University, Rajkot 360003, India
5 Department of Information Technology, College of Computer and Information Sciences, Princess Nourah bint Abdulrahman University, P.O. Box 84428, Riyadh 11671, Saudi Arabia
6 Security Engineering Lab, Computer Science Department, Prince Sultan University, Riyadh 11586, Saudi Arabia
7 Department of Electronics and Electrical Communications Engineering, Faculty of Electronic Engineering, Menoufia University, Menouf 32952, Egypt
* Correspondence: sudipta.das1985@gmail.com

Abstract: In this article, very compact 2 × 2 and 4 × 4 MIMO (Multiple-Input and Multiple output) antennas are designed with the help of Characteristics Mode Analysis to enhance isolation between the elements for UWB applications. The proposed antennas are designed with Characteristic Mode Analysis (CMA) to gain physical insight and also to analyze the dominant mode. To improve isolation and minimize the mutual coupling between radiating elements, elliptical shaped stubs are used. The dimensions of the 2 × 2 and 4 × 4 MIMO antennas are $0.29\lambda_0 \times 0.21\lambda_0$ (28×20 mm^2) and $0.29\lambda_0 \times 0.42\lambda_0$ (28×40 mm^2), respectively. These antennas cover the (3.1 GHz–13.75 GHz) UWB frequency band and maintain remarkable isolation of more than 25 dB for both 2 × 2 and 4 × 4 antennas. The impedance bandwidth of the proposed 4 × 4 MIMO antenna is 126.40% from 3.1 GHz to 13.75 GHz, including X-Band and ITU bands. The proposed 4 × 4 antenna has good radiation efficiency, with a value of more than 92.5%. The envelope correlation coefficient (ECC), diversity gain (DG), mean effective gain (MEG), and channel capacity loss (CCL) matrices of the 4 × 4 antenna are simulated and tested. The corresponding values are 0.0045, 9.982, −3.1 dB, and 0.39, respectively. The simulated results are validated with measured results and favorable agreements for both the 2 × 2 and 4 × 4 UWB-MIMO antennas.

Keywords: CMA; decoupling stub; ECC; isolation; UWB-MIMO

Citation: Suresh, A.C.; Reddy, T.S.; Madhav, B.T.P.; Das, S.; Lavadiya, S.; Algarni, A.D.; El-Shafai, W. Investigations on Stub-Based UWB-MIMO Antennas to Enhance Isolation Using Characteristic Mode Analysis. *Micromachines* **2022**, *13*, 2088. https://doi.org/10.3390/mi13122088

Academic Editors: Trushit Upadhyay and Hari Shankar Singh

Received: 29 October 2022
Accepted: 22 November 2022
Published: 27 November 2022

Publisher's Note: MDPI stays neutral with regard to jurisdictional claims in published maps and institutional affiliations.

Copyright: © 2022 by the authors. Licensee MDPI, Basel, Switzerland. This article is an open access article distributed under the terms and conditions of the Creative Commons Attribution (CC BY) license (https://creativecommons.org/licenses/by/4.0/).

1. Introduction

The release of the unlicensed frequency spectrum from 3.1 GHz–10.6 GHz (UWB) by the Federal Communications Commission resulted in significant changes in wireless communications, including high-speed transmission, data rate, and higher security. But the problems faced by UWB are fading, multipath environments, and low radiating powers of less than 41.3 dB/MHz [1]. These drawbacks of UWB technology reduce its performance. In the future, high data rates will be required with more than 100 Gbps of bandwidth to meet all wireless communications applications. The reliable technology called MIMO (Multiple Input and Multiple Output) increases the channel capacity, resulting in high data rates and high speeds of data transmission. The MIMO maximizes the reduction in multiple paths and fading problems so that the combined UWB-MIMO technology attains better link quality and higher data rates. However, the challenge faced by UWB-MIMO is the size of the antennas. Several MIMO elements are integrated into a small space, producing high mutual coupling and correlation problems. Therefore, several methods have been adopted to enhance isolation and reduce mutual couplings, including introducing slots, placing stubs, and Electromagnetic Band-Gap (EBG) in between MIMO elements and also placing conducting materials called parasitic materials in between MIMO elements. Spatial and pattern diversity techniques are also used to improve isolation in MIMO elements [2]. The periodical arrangements of Complimentary Splint Ring Resonator (CSRR) metal strips or capacitive gaps perform filtering action and enhance isolation between MIMO elements [3].

In the UWB antenna with dimensions 100 × 150 mm^2, a metasurface-based decoupling method (MDM) was used, and antenna elements were composed of the metasurface's superstrate to achieve isolation greater than 25 dB [4]. In the Fractal MIMO antenna, an L-shaped inverted slot was used to reduce the mutual coupling up to 34 dB [5,6]. The half-hexagonal-shaped monopoles included grounded circular rings along with grounded stubs that improved the isolation by 20 dB [7]. Several decoupling methods are used to maximize isolation [8–12]. A very high-compact four-port vertical polarized UWB MIMO antenna covered a 2.2 GHz–12.3 GHz band with high isolation of more than 15 dB. A Specific interlocking decoupling method was used to achieve low mutual coupling [13]. A very sophisticated mutual coupling technique is called "neutralizing lines," which were added into 2 × 2 and 4 × 4 very compact MIMO elements to enhance isolation, −22 dB, and achieve a low CCL of less than 0.29 bits/Hz [14]. Pentagonal MIMO antenna structures were investigated to achieve high isolation of more than 28 dB by inserting slots and plus-shaped parasitic structure between the patch elements [15].

A miniaturized 4 × 4 UWB MIMO antenna has been designed with dimensions of 40 × 40 × 1.567 mm^3. This antenna is very simple in structure and covers UWB (3.1 GHz-10.6 GHz) and ITU bands (3.1 GHz–13.25 GHz), with high isolation of more than 15 dB, and the achieved radiation efficiency is 89% [16]. A rectangular microstrip stub with defected ground plane is employed to increase the isolation >15 dBfor the reported dual band MIMO antenna [17]. However, the above work provides only partial isolation improvement techniques, because the design process lacks a systematic approach, so it does not meet the quality requirement of antenna design. A systematic design approach is required to understand the physical insight of the antenna structure, which gives immense knowledge of the physical structure to antenna designers and allows a good design process of the antenna in less time with more sophisticated results. The generated currents and scattered fields from perfect electric conductors are explained by CMC [18,19].

The mode currents are the sum of orthogonal currents from PEC bodies, and modes have orthogonal characteristics across the surface [20]. The various advantages of CMA are used to design sophisticated antennas in a systematic design approach. The CMA metrics and modal significance provides the bandwidth, coupling capability, and insight into the exciting mode of the radiating element without exciting the feed [21]. The performance of multiple antennas in a systematic design approach is achieved by Characteristic Mode Analysis [22]. Two metal strips were combined to form the chassis to generate two characteristic modes with wideband properties, and to achieve high isolation [23,24]. The diversity radiation pattern and high isolation are achieved by two monopole antennas. One radiator is the compact quarter loop and the second one is a circular planar monopole. Both are ultra-wide antennas located at opposite faces of the rectangular ground plane of MIMO, and both elements are excited in different modes [25]. By using the symmetry of the etched slots, the required isolation of more than 32 dB is achieved between antenna elements [26]. A 4 × 4 reconfigurable UWB MIMO antenna is designed to achieve isolation greater than 20 dB in an entire band of 3 GHz-11 GHz. This antenna shows some band rejection characteristics due to the placing of the PIN diode. Different geometry shapes and novel decoupling techniques were also used to achieve high isolation, good radiation performance, and improved diversity [27–30]

The majority of the preceding work employed orthogonal modes and defected ground structures to improve isolation in MIMO antennas, and some antennas employed space diversity techniques. The above works need a more systematic design approach.

In this work, a new systematic methodology is proposed to achieve improved isolation in MIMO antenna using Characteristic Mode Analysis. The elements in the proposed antenna, 2 × 2 and 4 × 4 UWB-MIMO, are placed symmetrically over the FR-4 substrate to achieve high compactness.

Novelty and Contributions

The main important points focused on while designing the 2 × 2 and 4 × 4 MIMO antennas are:

1. The proposed compact 2 × 2 and 4 × 4 UWB-MIMO antennas achieved −25 dB isolation among elements using Characteristic Mode Analysis (CMA).
2. The designed antennas not only cover UWB but also cover the X and ITU bands.
3. The decoupling I-shaped stub is placed at ground level to enhance the isolation and overall performance of the MIMO antennas.
4. A step-by-step systematic designing approach using CMA methodology is used to achieve good diversity characteristics in operating frequency range.
5. Both 2 × 2 and 4 × 4 antennas achieve good gain and radiation efficiencies.
6. The designed 2 × 2 UWB-MIMO achieved a peak gain and radiation efficiency of 4.2 dB and 82.5%, respectively. Therefore, there is a scope to enhance the above said parameters by changing the 2 × 2 configuration into a 4 × 4 MIMO arrangement.
7. The proposed 4 × 4 UWB-MIMO with CMA achieved gain and radiation efficiency up to 4.8 dB and 92.2%, respectively. The ECC of 2 × 2 is also reduced from 0.005 to 0.0045 by designing 4 × 4 MIMO antenna.

The rest of the paper is organized as follows. The detailed design of 2 × 2 with CMA is explained in Section 2. Section 3 explains the result of 2 × 2 MIMO antenna. Section 4 is organized to explain the 4 × 4 UWB-MIMO antenna. Performance characteristics and results of 4 × 4 antenna are analyzed in Section 5, and in Section 6 comparisons with existing models are discussed. Finally, this paper ends with conclusions in Section 7.

2. Antenna Design

2.1. UWB-MIMO Using CMA

The designed 2 × 2 antenna is composed of two symmetrical circular radiators at the top, each having rectangular discs that behave in a similar way to radiating elements. The ground plane is composed of two elements: (i) partial ground plane and (ii) elliptical shape decoupling stub.

The overall dimensions of the proposed 2 × 2 antenna are $0.29\lambda_0 \times 0.21\lambda_0$ (28×20 mm^2) printed on an FR-4 substrate, with $\tan\delta = 0.002$ and $\varepsilon_r = 4.3$, as depicted in Figure 1. To improve low-frequency isolation, this proposed antenna uses a circular patch and a rectangular disc. The designed parameters with dimensions are listed in Table 1.

Table 1. The designed parameters (Units in mm).

L1	L2	L3	d1	W1	W2
2	7	12.5	17.5	4	1.2
L4	r1	r2	r3	r4	r5
3.4	3	2	3	2	7
r6	Ws	W_L	Wh	hp	L5
1.735	20	20 & 40	1.4	0.035	8
L6	r				
8	3.6				

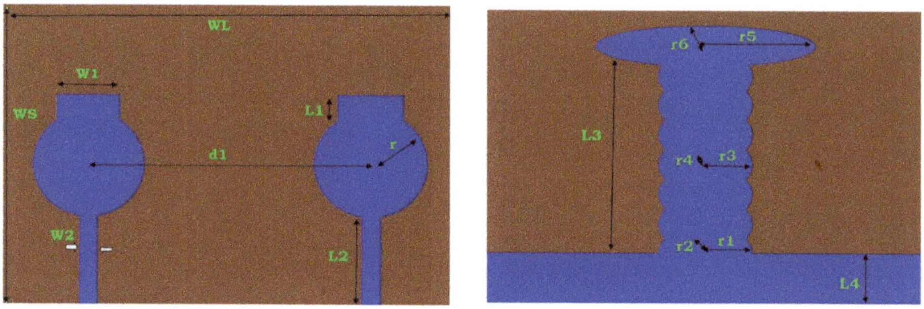

(a) (b)

Figure 1. The dimensions of the proposed high-compact 2 × 2 UWB MIMO antenna. (**a**) Top view, (**b**) Bottom view.

The radius (r) of 2 × 2 circular Patch radiator [31] is determined by using the following expressions:

$$r = \frac{F}{\sqrt{1 + \frac{2h}{\pi \epsilon_r F}\left[\ln\left(\frac{\pi F}{2h}\right) + 1.7726\right]}} \tag{1}$$

where F is given by:

$$F = \frac{8.791 \times 10^9}{f_r\sqrt{\epsilon_r}} \tag{2}$$

The effective radius results from the fringing field spreading from the patch border to the ground plane. The fringing field around the circular patch may increase the radius of a circle. Therefore, the effective radius (r_{eff}) is determined as shown below:

$$r_{eff} = \frac{1.8412 \times c}{2\pi f_r\sqrt{\epsilon_r}} \tag{3}$$

where c is velocity of light in vacuum = 3×10^{11} mm/s, and f_r is resonant frequency. The dielectric material FR-4 is used, its height is h = 1.6 mm, and its operating frequency is between 3.1 GHz and 13.75 GHz. This proposed antenna well resonates in the UWB range. The specifications are ϵ_r = 4.3 and h = 0.16 cm, and f_r = 7.5 GHz is used for calculating the radius of the patch.

$$r = \frac{F}{\sqrt{1 + \frac{2h}{\pi \epsilon_r F}\left[\ln\left(\frac{\pi F}{2h}\right) + 1.7726\right]}} = \frac{0.6}{\sqrt{1 + \frac{2\times 0.16}{\pi \times 4.3 \times 0.6}\left[\ln\left(\frac{\pi \times 0.6}{2\times 0.16}\right) + 1.7726\right]}} = 3.6 \text{ mm}$$

The radius (r) = 3.6 mm is suitable for resonating the antenna in UWB band.

The reflection coefficient of the single radiator is initially analyzed by changing the horizontal stub (W1) from 3.5 mm to 4.25 mm. The corresponding reflection coefficient and isolation are depicted in Figure 2a,b. The same procedure is repeated by varying the values of the Length of the rectangular disc (L1) and the distance between two radiating elements (d1), and its corresponding reflection coefficients are depicted in Figure 2c,d.

The horizontal disc's width (W1) was adjusted from 3.5 mm to 4.25 mm and the current distributions were simultaneously observed. The current distributions were excellent at low frequencies when W1 = 3.5 mm, but the UWB range was not covered. The current distributions were excellent at high frequencies (>10.6 GHz) at W1 = 4.25 mm. However, at lower frequencies and in the UWB range they were not appreciable. The current distributions shifted from the low-frequency to the middle-frequency area at W1 = 4 mm. Thus, at W1 = 4 mm, the designed MIMO effectively spanned the impedance bandwidth.

Therefore, the antenna provides good impedance at W1 = 4 mm and, as a result, good current distribution in the UWB. The horizontal disc's length (L1) was adjusted from 1.5 to 2.25 mm, and the current distribution was observed. The current distribution was excellent at lower frequencies when L1 was 1.5 mm. However, it was not impressive in the UWB and higher frequency regions. When the L1 was changed from 1.5 mm to 2.25 mm, the current distribution over the antenna surface shifted in the middle portion from the low-frequency region to the high-frequency region and did not cover the UWB range. The maximum current distributions shifted from the low-frequency to the middle-frequency area at L1 = 2 mm, so that at L1 = 2 mm, this MIMO adequately covered the good impedance bandwidth's entire UWB region.

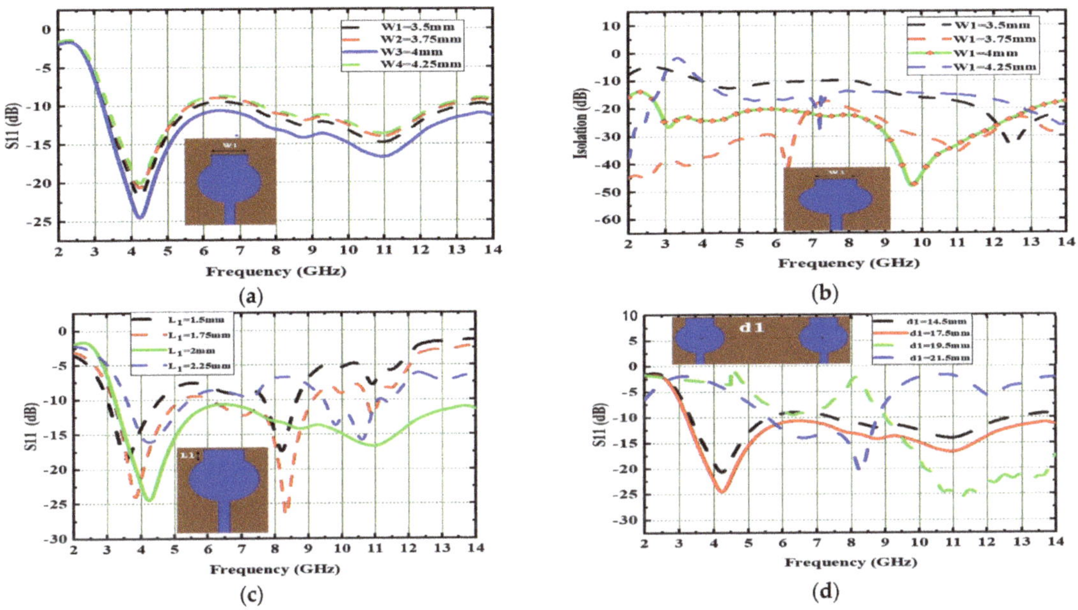

Figure 2. Parametric variations of single radiator w.r.t. (**a**) Width of the of the horizontal disc W1, (**b**) Isolation due to variations in W1, (**c**) Variations in L1 (Length of rectangular disc L1), (**d**) Variations in d1 (Distance between two radiators d1).

The isolation between the MIMO elements has to be maintained, otherwise the interaction between the elements would increase and the MIMO diversity characteristics may not meet. The space diversity technique can be used to improve the isolation. The current distribution interaction between MIMO elements can be reduced by using additional space between the MIMO elements, but this technique needs more space. The distance between elements (d1) in this MIMO design ranged from 14.5 mm to 21.5 mm. Low mutual coupling between the MIMO components was not achieved for the element spacing of d1 = 14.5 mm. The mutual coupling between the elements is weak at d1 = 21.5 mm, but the antenna size increases. The selected value of d1 = 17.5 mm obtained good isolation among the antenna elements for the MIMO's compactness.

Antenna 0 (A#0), Antenna 1(A#1), Antenna 2(A#2), and Antenna 3(A#3) are the four stages of the design evolution process. The proposed highly compacted 2 × 2 antenna can be transformed into a 4 × 4 MIMO antenna without the need for additional design methods.

The design procedure of 2 × 2 is performed by Characteristic Mode Analysis. The physical significance of the antenna is analyzed by eigenvector, characteristic angle, and modal significance, which is referred to as CMC performance matrices. The proposed antenna design and its performance improved by CMA without applying any excitation. After that, these results were cross-verified by applying feed excitation, which is called time-domain analysis. By applying the systematic design approach step by step, the overall bandwidth is improved and MIMO performance metrics are also improved. The UWB-MIMO antenna was designed with a systematic approach using CMA in Computer Simulation Tool (CST, Ver. 2018) Multilayer solver or Integral Equation solver. The Evolution of the step by step design process was performed using a systematic method and is depicted in Figure 3.

(a) (b) (c) (d)

Figure 3. Evolution of 2 × 2 UWB MIMO antenna. (**a**) A#0, (**b**) A#1, (**c**) A#2, (**d**) A#3.

Generally, the modal significance of these four antennas, A#0, A#1, A#2, and A#3, are evaluated without applying any feed using CMA, as shown in Figure 4. The same four antennas, A#0, A#1, A#2, and A#3, were analyzed using time-domain analysis by applying feed and excitation; the associated S-parameters are depicted in Figure 5. Initially, the high-compact 2 × 2 antenna was constructed with two circular patch monopoles and was composed with rectangular discs without a ground plane, referred to as A#0. The A#0 was analyzed by generating characteristic modes. In A#0, almost five characteristic modes were generated to analyze physical insight into the structure of the antenna, and associated modal significance are depicted in Figure 4a. The characteristics modes 1, 2, 4, and 5 contribute to achieving a higher bandwidth. The CM3 did not contribute anything to achieve wideband. The modal significances of A#1, A#2, and A#3 are depicted in Figure 4b–d. The associated S-parameters of A#0, A#1, A#2, and A#3 are shown in Figure 5a–d.

The antenna offers low isolation at low frequencies, but it is also quite good at the remaining frequencies in the entire UWB spectrum, as observed from its S-parameters. In the design evolution of the high-compact 2 × 2 MIMO antenna, the fundamental Characteristic Mode Current (CMC) changes the direction of interaction with MIMO elements, as depicted in Figure 6.

Figure 6 explains the CMs of CMC in A#0, A#1, A#2, and A#3, respectively. The fundamental CM1 of CMC interacts with one monopole rather than the ground plane to the monopole because A#0 is not in contact with the ground. Hence, greater interference is generated between the two radiating monopoles in A#0, which consequently increases the mutual coupling. The corresponding current distributions of A#0 in CM1is depicted in Figure 6a. When CMA is applied to the A#0, it generates various CMs, each with its own modal significance, eigenvalues, and characteristic angle. The eigenvalue represents the effectiveness of a mode, while the modal significance represents relative bandwidth. The characteristic angle denotes the current distribution's compactness at that frequency.

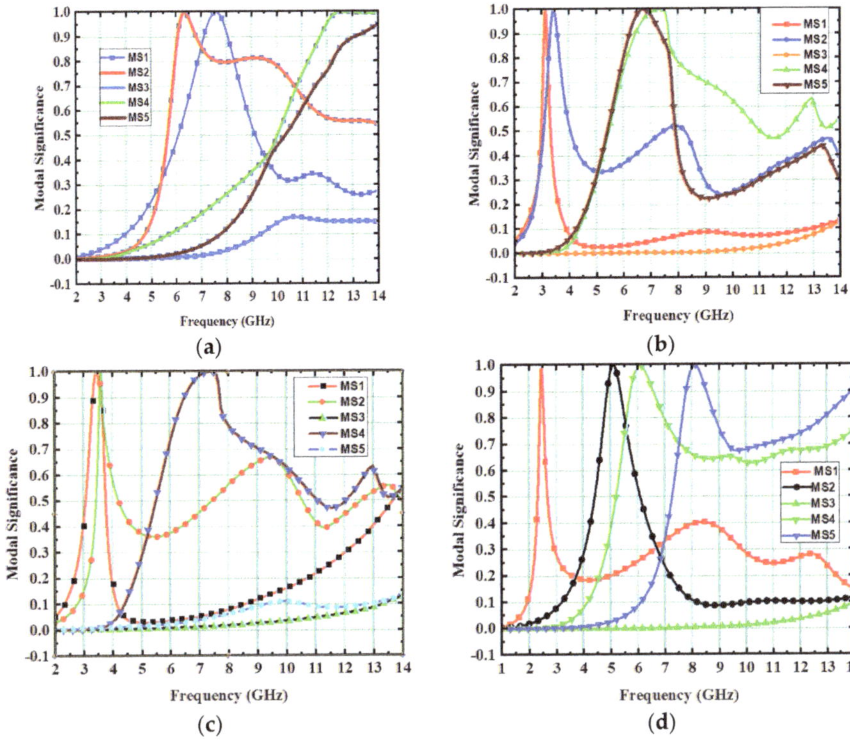

Figure 4. The modal significance of 2 × 2 UWB MIMO antenna. (**a**) A#0, (**b**) A#1, (**c**) A#2, (**d**) A#3.

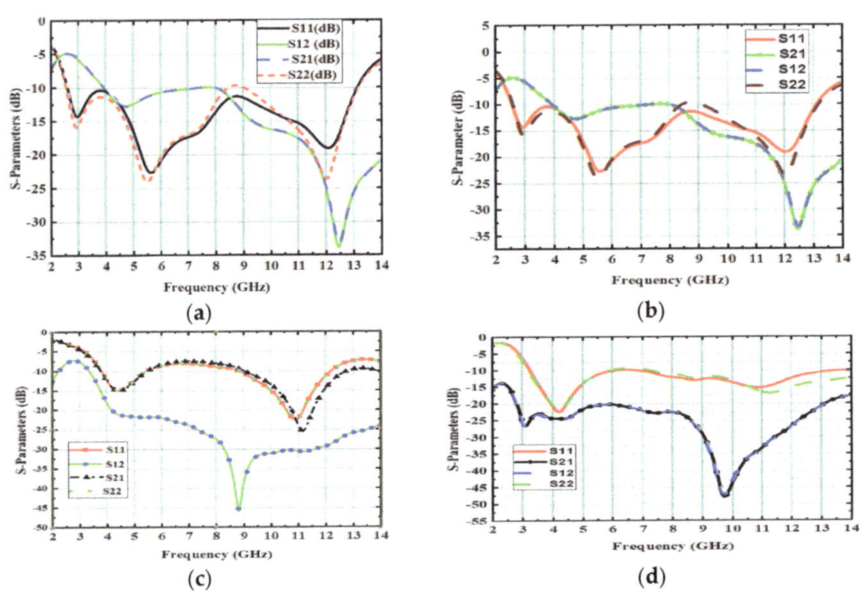

Figure 5. S-parameter of proposed 2 × 2 UWB MIMO antenna. (**a**) A#0, (**b**) A#1, (**c**) A#2, (**d**) A#3.

(a) (b) (c) (d)

Figure 6. The fundamental CMC effect in the evolution of the 2 × 2 UWB-MIMO.

When compared to other CMs, the modal significance closest to unity ($M_s = 1$) indicates that the mode resonates and is most effective at that frequency. The eigenvalue reaches zero ($\lambda_n = 0$) when that mode of operation is resonated at that frequency. The characteristic angle where it crosses 180° ($\theta c = 180°$) is where the corresponding frequency resonates at that frequency.

In comparison to other modes, modal significance reaches one at A#0, characteristic angle crosses 180°, and eigenvalue is zero for CM1, indicating that this CM1 resonates at this frequency and is more effective in the low-frequency range. The A#1 is composed of partial ground, and its current densities are at different frequencies in the entire operating frequency spectrum, as depicted in Figure 7.

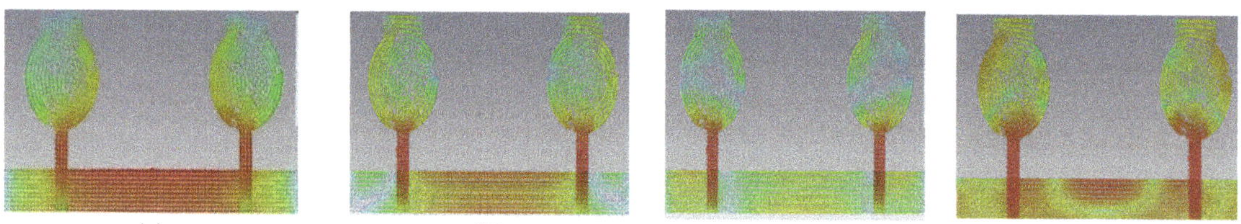

(a) (b) (c) (d)

Figure 7. The A#1 surface current densities at (**a**) 3.8 GHz, (**b**) 6.8 GHz, (**c**) 9.8 GHz, (**d**) 12.8 GHz.

A#1 is obtained by adding the ground plane to A#0. Therefore, all CMs are moved toward the lower frequency region, except CM1. Most of the current distributions and currents are now moving from the monopole to the ground plane of the MIMO, i.e., low interference is achieved between one monopole radiator to another monopole radiator. The modal significance of A#1 is depicted in Figure 4b. The corresponding A#1 S-parameters are shown in Figure 5b. The radiation properties at high frequencies of A#1 are enhanced compared to A#0, i.e., the bandwidth is increasing towards high frequencies. However, due to the high interacting currents between radiating elements, good isolation is not achieved at low frequencies.

The design evolution process of the proposed antenna is executed via four (4) different design steps. Wideband frequency response was achieved by adjusting the CMs in the high-frequency region, at the high-frequency band, but a good impedance bandwidth was not achieved. Therefore, for this purpose, an I-shaped stub has been embedded in between the radiators to enhance the isolation of the elements. The design process of the I-shaped stub is explained in the following sections. The next stage of evolution is A#2.

2.2. *Effect of Decoupling Stub*

The various types of decoupling techniques, such as F-shaped stubs, Y-shaped stubs, and T-shaped stubs, are used to reduce mutual coupling [32–34]. The decoupling stubs are placed in between radiating elements to enhance isolation in the MIMO antenna. In this design evolution process of antenna, an I-shaped decoupling stub is combined with A#1 to form A#2. By adding an I-shaped stub to A#1, further mutual coupling is reduced. The dimensions of an I-shape stub are listed in Table 1 and are formed by stacking seven ellipses one on top of the other. The A#2 is analyzed with CMA without applying excitation. Due to the insertion of the I-shaped stub in A#2, all modes are shifted from the high-frequency band to the mid-frequency band, except CM1, hence increasing the bandwidth and isolation between radiating elements. The respective modal significance is depicted in Figure 4c.

The current distributions in A#2 interact with the monopoles, I–shaped stub, and the ground plane. Earlier, the scattered currents interacted from monopoles to monopoles, and monopoles to the ground. Now the currents are moving from stubs to the ground and from monopoles to the ground, so mutual coupling is reduced between radiating elements. The current distributions of A#2 at different frequencies are depicted in Figure 8.

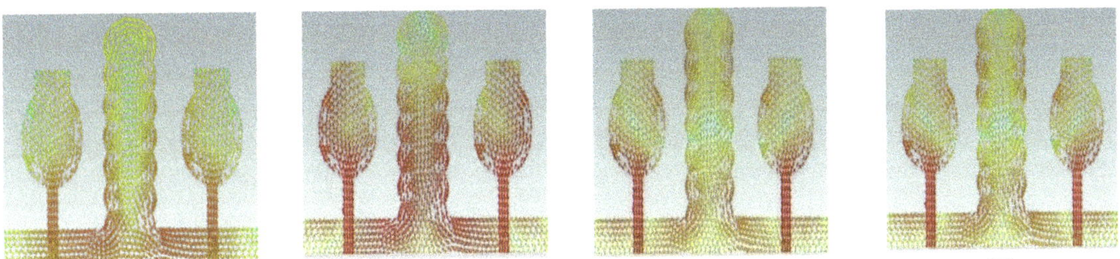

(a) **(b)** **(c)** **(d)**

Figure 8. The characteristic currents of CM1in A#2 at (**a**) 3 GHz, (**b**) 6 GHz, (**c**) 9 GHz, and (**d**) 12 GHz.

The A#2 is analyzed by applying feed with excitation in time-domain analysis. The respective S-parameters are shown in Figure 5c. Figure 5c explains that the isolation of the monopole radiators is increased. However, mutual interaction between the elements is still observed at some frequencies that will degrade the performance matrices of MIMO, such as CCL, MEG, ECC, and DG. By placing I-shaped decoupling stubs in MIMO, they reduce the mutual coupling, but not to a higher extent.

Therefore, we need to enhance further isolation. For that purpose, we added an elliptical patch at the top of the I-shaped stub, and it becomes a T-shaped stub. The resultant antenna is called A#3. Characteristic Mode Analysis is applied to A#3 to generate the respective five characteristic modes. By adding an elliptical patch to the I–shaped stub, most of the characteristic modes are shifted from the high-frequency region to the mid-band frequency spectrum [35], which results in an increase in the bandwidth over the UWB frequency spectrum. The corresponding modal significances are shown in Figure 4d. The time-domain analysis is applied to A#3 with feed and excitation. The respective S-parameters are observed and depicted in Figure 5d. The scattering fields and current distributions are moving among radiating elements, stubs, and the ground plane, instead of moving from radiator to radiator. Hence, high isolation is achieved in this antenna.

The time-domain analysis is applied to A#3 with feed and excitation. The respective S-parameters are depicted in Figure 5d. By integrating elliptical patch on to an I-shaped decoupling stub, the isolation of the MIMO antenna is improved, and good performance matrices are achieved. At various frequencies, the A#3 characteristic currents and current densities are simulated and shown in Figures 9 and 10, respectively [36–38].

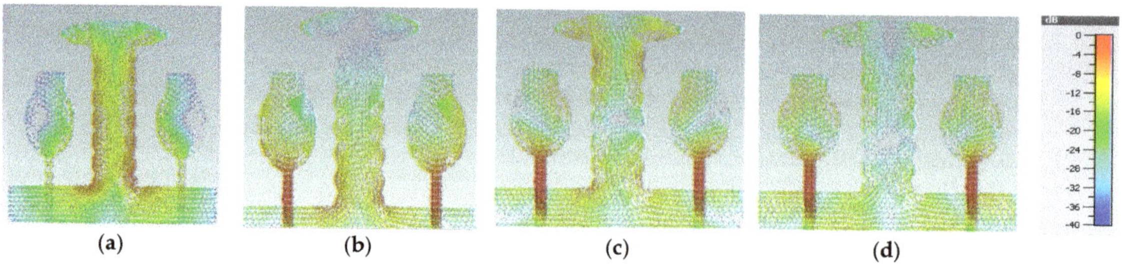

Figure 9. The effect of characteristic currents due to CM1 in A#3 at (**a**) 3 GHz, (**b**) 4 GHz, (**c**) 5 GHz, and (**d**) 6 GHz.

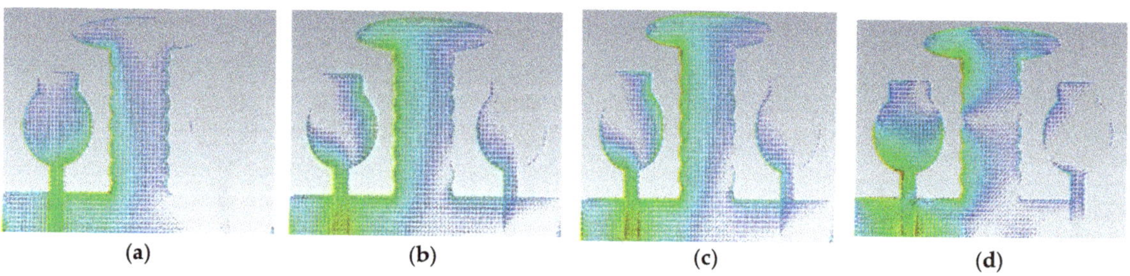

Figure 10. Current densities due to CM1 in A#3 at (**a**) 3 GHz, (**b**) 4 GHz, (**c**) 5 GHz, and (**d**) 6 GHz.

3. Results of High-Compact 2 × 2 UWB-MIMO Antenna

The performance of the designed 2 × 2 antenna was measured and found good isolation from 3.1 GHz to 13.25 GHz of the entire frequency spectrum, with more than 23 dB isolation and a gain of 4.2 dB. The radiation efficiency and fractional impedance bandwidth is 82.5% and 124.15%, respectively. The fabricated compact UWB MIMO 2 × 2 is depicted in Figure 11. The return losses of an antenna are found in $S_{11} \leq 10$ dB of the entire UWB spectrum. Isolation is less than 23 dB and is depicted in Figure 12. Figure 13 represents a comparison of tested and simulated values of S_{11} and S_{22}. Figure 14 represented the comparisons of simulated and tested values of isolation between the elements of 2 × 2 UWB-MIMO Antenna.

The overall performance of MIMO depends on its gain and efficiency. By using a sophisticated step by step design procedure, this antenna achieved 4.2 dB gain and 82.5% radiation efficiency, as depicted in Figure 15. The measurement of gain and its setup is depicted in Figure 15c. The key performance of MIMO antennas is diversity performance, and antenna designers focus on diversity characteristic parameters while designing UWB-MIMO antennas to meet the challenges in wireless communication. One of the most significant parameters of the diversity parameter is the ECC, which shows how the radiating elements interact with each other when placed very close together in MIMO. The respective ECC is depicted in Figure 16b, and its value is 0.005. Diversity gain is measured, and its value is 9.89, as shown in Figure 16a. The E-pattern and H-pattern are tested, and simulated results are depicted in Figure 17 at frequencies of 4.5 GHz and 6.8 GHz, respectively.

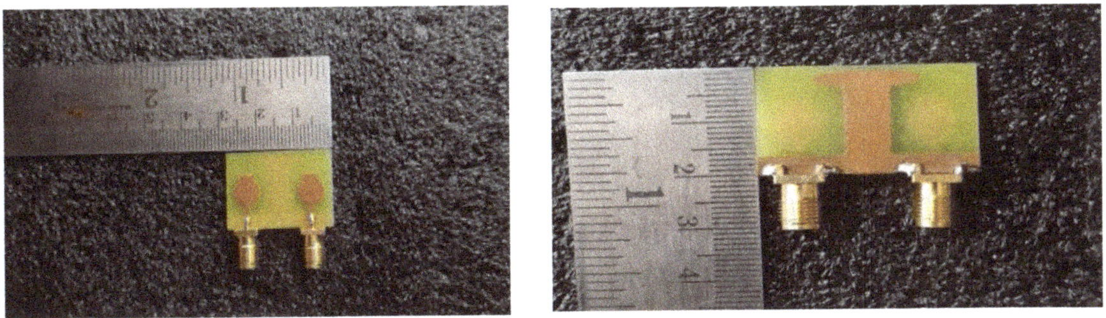

(**a**) (**b**)

Figure 11. The Proposed 2 × 2 UWB-MIMO fabricated antenna (**a**) Front View (**b**) Back View.

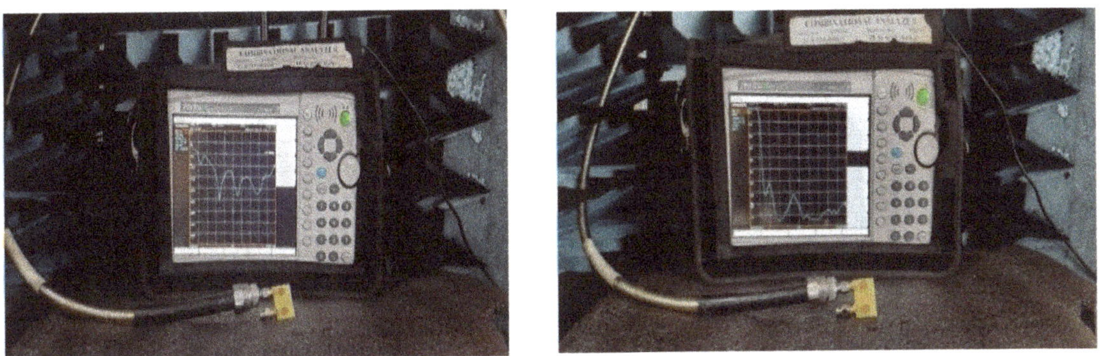

(**a**) (**b**)

Figure 12. Prototype of the proposed 2 × 2 UWB-MIMO antenna. (**a**) Return Loss, (**b**) Isolation.

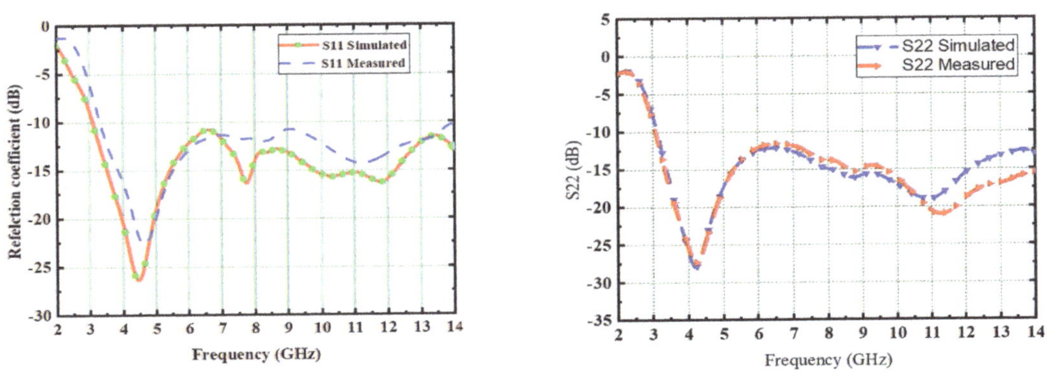

(**a**) (**b**)

Figure 13. Comparisons of simulated and measured (**a**) Reflection coefficient (S11) (**b**) S22.

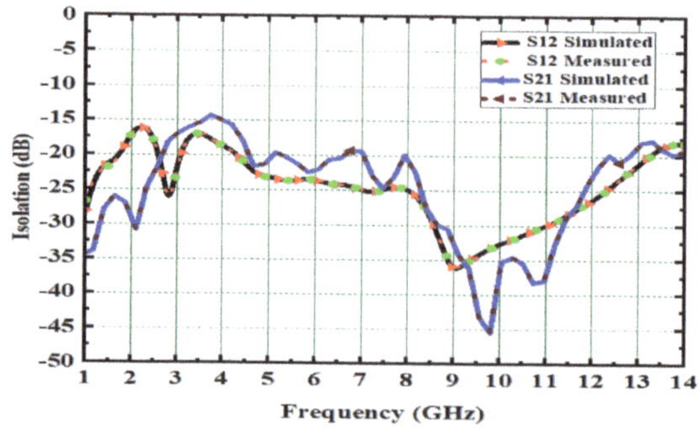

Figure 14. Isolation in 2 × 2 UWB-MIMO Antenna.

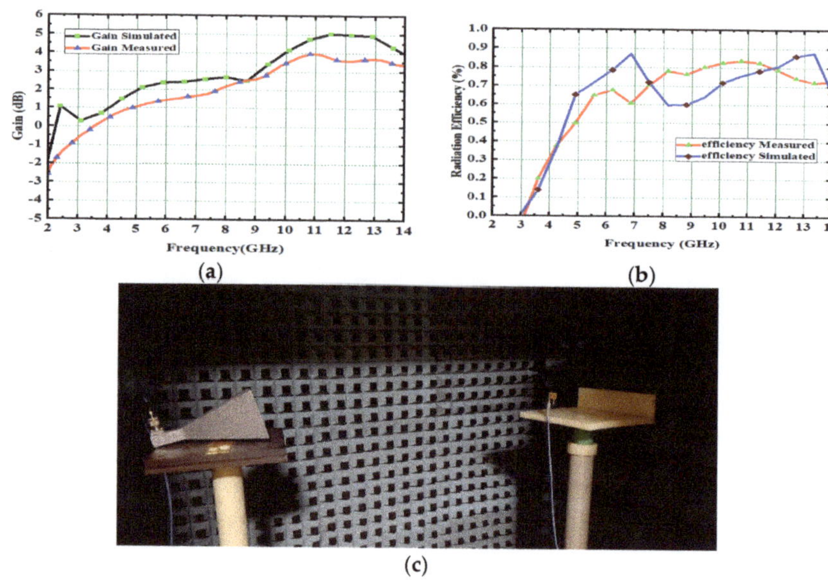

Figure 15. (**a**) Gains and (**b**) Radiation efficiency, (**c**) Gain measurement setup.

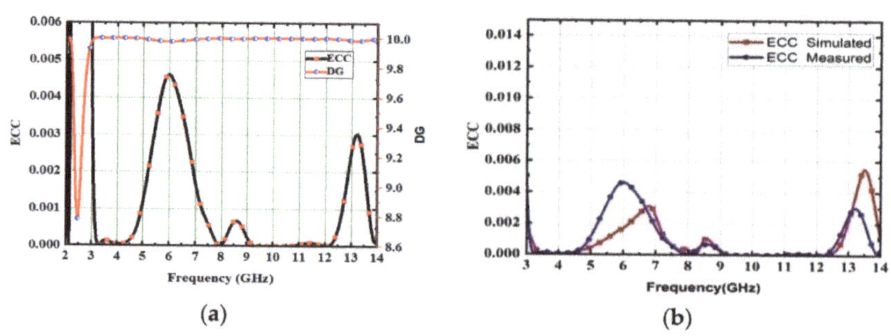

Figure 16. (**a**) DG with ECC, (**b**) ECC Measured.

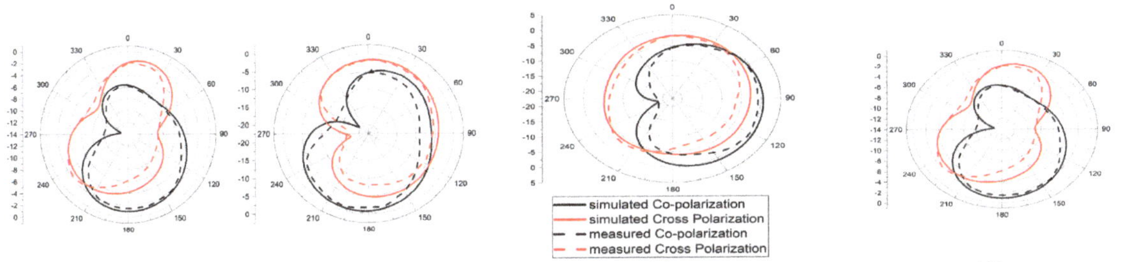

(a) (b) (c) (d)

Figure 17. The radiation Pattern of E-Plane (**a**) 4.5 GHz and (**b**) 6.8 GHz, and H-plane at (**c**) 4.5 GHz and (**d**) 6.8 GHz.

Still, there is a great need to achieve high isolation and ECC for reliable wireless communication. Therefore, we can extend this 2 × 2 UWB-MIMO antenna into a 4 × 4 UWB-MIMO antenna to achieve good diversity performance using Characteristic Mode Analysis.

4. 4 × 4. UWB-MIMO Antenna Design

The designed compact 2 × 2 UWB-MIMO antenna has dimensions of 0.29 λ_0 × 0.21 λ_0 (28 × 20 mm^2), and it is extended into a 4 × 4 high-compact UWB-MIMO antenna with dimensions of 0.290 × 0.420 (28 × 40 mm^2) printed on FR-4 having ε_r = 4.3 with tanδ = 0.002.

This 4 × 4 UWB-MIMO design process is performed in stages denoted as A#0, A#1, A#2, and A#3, respectively. The 4 × 4 UWB-MIMO systematic design process is depicted in Figure 18. Increasing the elements in MIMO will increase diversity performance. The antenna is designed without the ground plane and applies characteristic modes to observe its modal significance, as seen in the corresponding graphs shown in Figure 19a. Figure 20a explains the S-parameters of the above A#0. The CMs are good at high frequencies but not at low frequencies. A ground plane is added to A#0, along with a decoupling stub to form A#1, A#2, and A#3; the corresponding modal significance and S-parameters are depicted in Figures 19b–d and 20b–d.

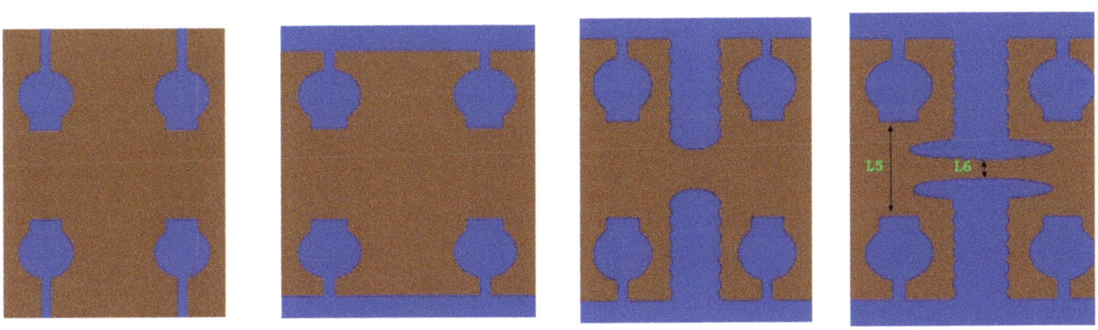

(a) (b) (c) (d)

Figure 18. The Evolution of 4 × 4 UWB MIMO antenna design. (**a**) A#0, (**b**) A#1, (**c**) A#2, (**d**) A#3.

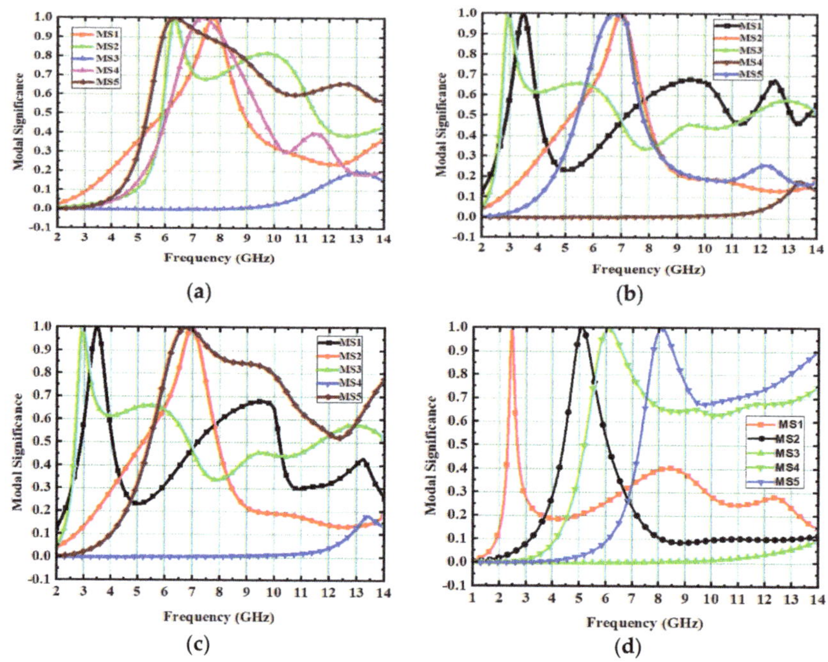

Figure 19. Modal significance of 4 × 4 antenna. (**a**) A#0, (**b**) A#1, (**c**) A#2, (**d**) A#3.

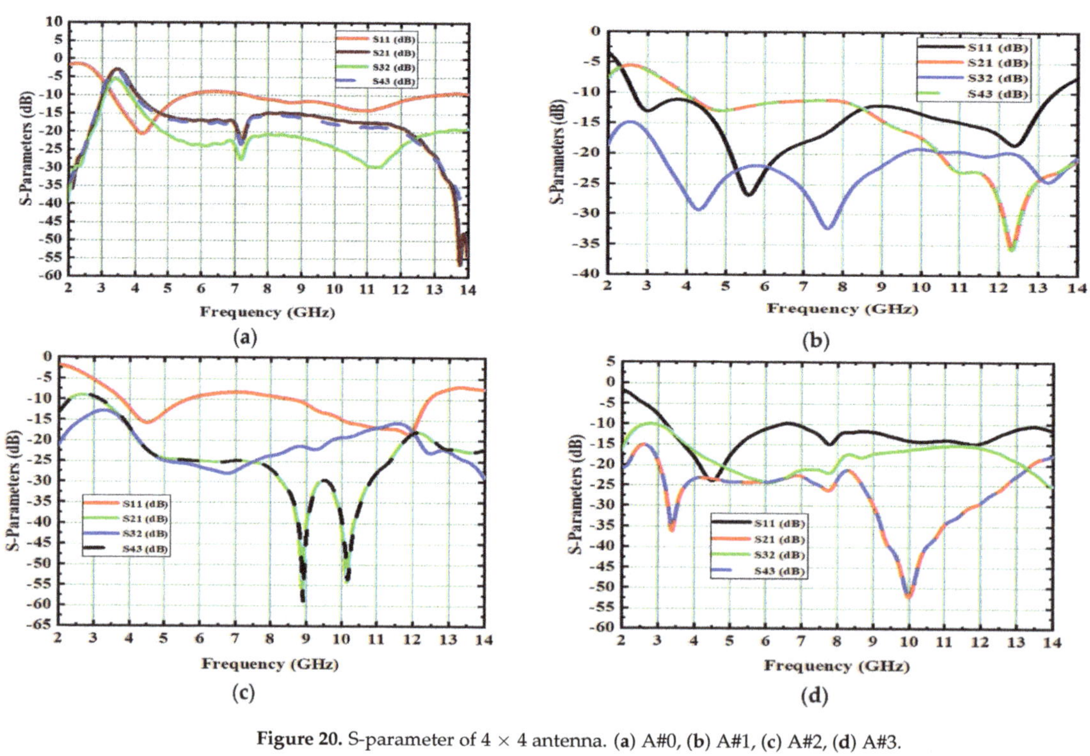

Figure 20. S-parameter of 4 × 4 antenna. (**a**) A#0, (**b**) A#1, (**c**) A#2, (**d**) A#3.

A#0 is designed without a ground plane, A#1 has a partial ground plane, A#2 is made up of an I-shaped stub, and A#3 is made up of an elliptical stub. The fundamental characteristics of currents and the distribution of CM1 are depicted in Figure 21. Adding stubs to a 4 × 4 antenna is similar to adding stubs to a 2 × 2 antenna. The surface current density and its distributions of A#1 at various frequencies are shown in Figure 22. By adding an I-shaped stub to A#1, it becomes A#2. The A#2 characteristic currents and distributions of CM1 are expressed at various frequencies, as depicted in Figure 23. For A#3, the characteristic current effect due to CM1 and current densities at different frequencies are depicted in Figures 24 and 25, respectively.

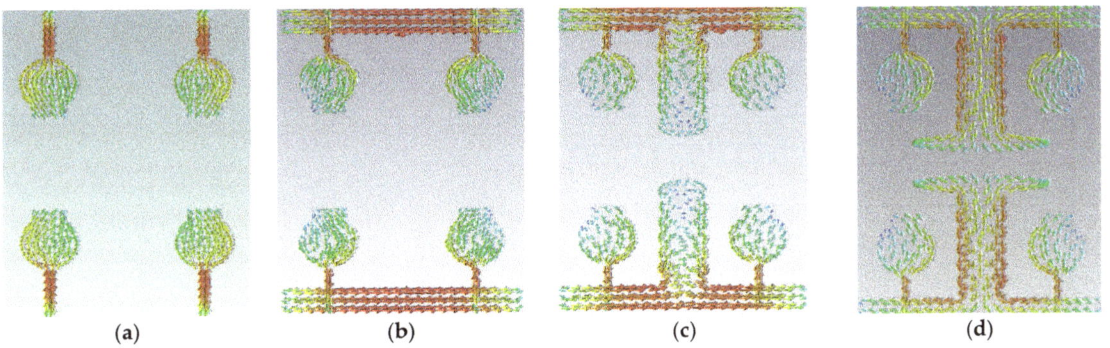

Figure 21. The characteristic mode currents in the design process of high-compact 4 × 4 UWB-MIMO antenna. Characteristic current effect due to CM1 of A#3 at (**a**) 3 GHz (**b**) 4 GHz (**c**) 5 GHz (**d**) 6 GHz.

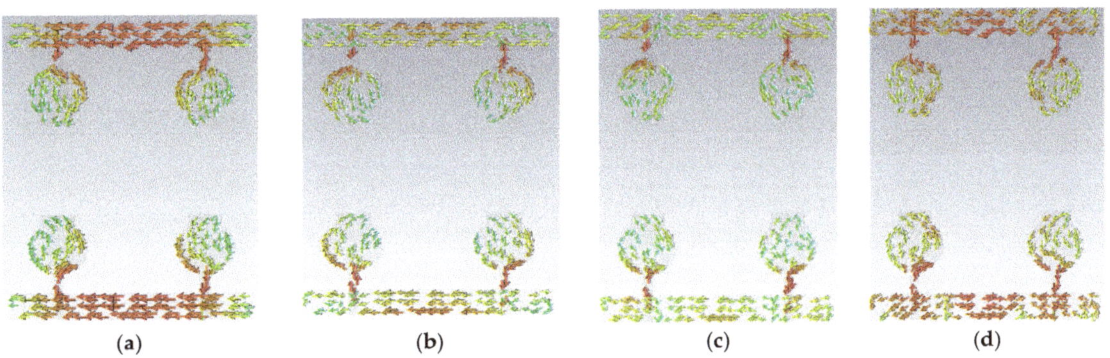

Figure 22. The surface current densities of A#1 at (**a**) 3.8 GHz, (**b**) 6.8 GHz, (**c**) 9.8 GHz, and (**d**) 12.8 GHz.

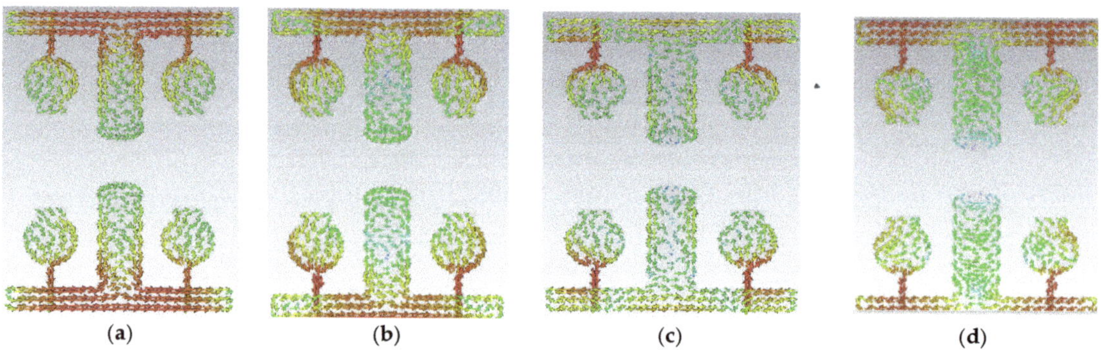

Figure 23. The effect of characteristic mode current in A#2 (**a**) 3 GHz (**b**) 6 GHz (**c**) 9 GHz (**d**) 12 GHz.

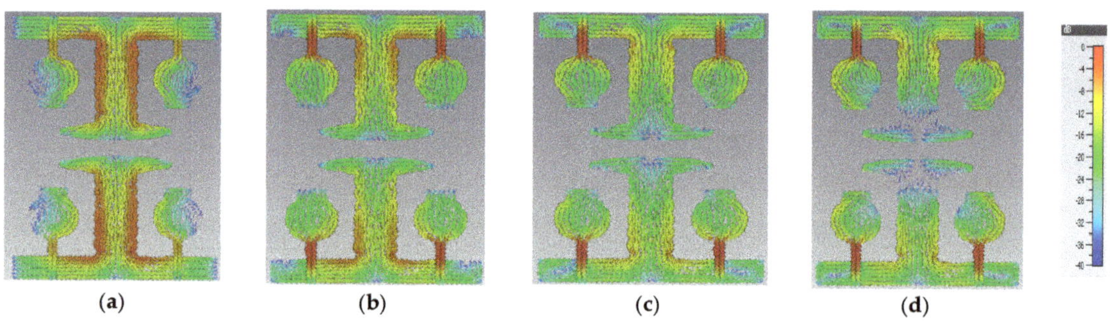

Figure 24. A#3 Characteristic current effect due to CM1 at (**a**) 3 GHz (**b**) 4 GHz (**c**) 5 GHz (**d**) 6 GHz.

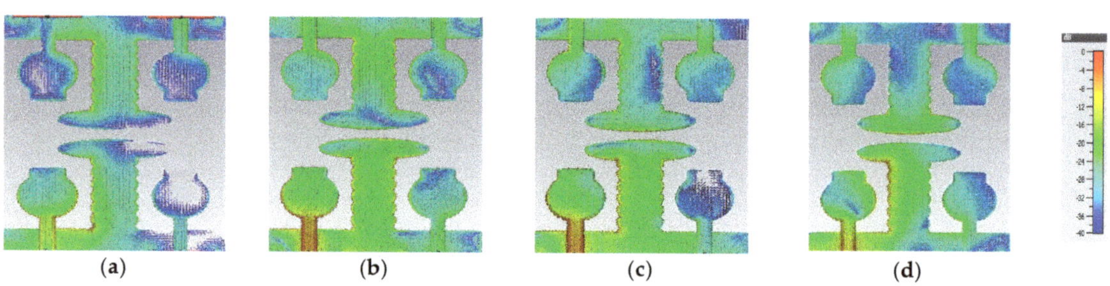

Figure 25. Surface current density of A#3. (**a**) 3 GHz, (**b**) 4 GHz, (**c**) 5 GHz, (**d**) 6 GHz.

5. Results of High-Compact 4 × 4 UWB-MIMO Antenna

The performance of the MIMO depends on its reflection coefficient and isolation among the MIMO elements. Good isolation is achieved through a systematic design approach called Characteristic Mode Analysis. The designed antenna covers 3.1 GHz–13.75 GHz at a bandwidth of 10.65 GHz. The prototype of 4 × 4 is depicted in Figure 26. The snapshot of the measured reflection coefficients and isolation using VNA are shown in Figure 27. The simulated values of S_{11} and S_{22} are compared with tested values, as depicted in Figure 28. Both the results are in acceptable agreement, as displayed and achieved with the help of CMA. The measured reflection coefficients and isolation are $\leq$10 dB and $\leq$25 dB, respectively.

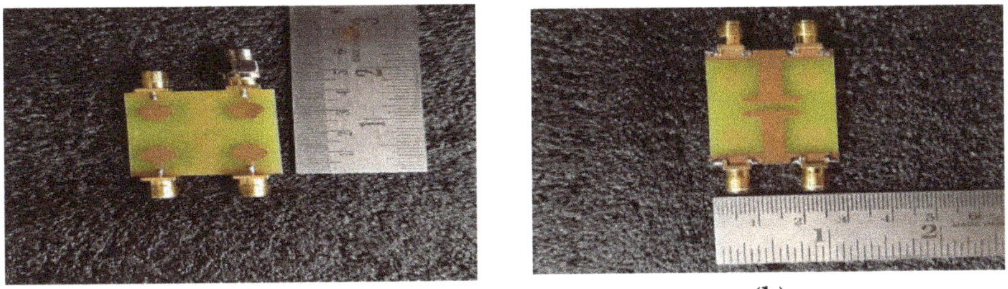

(**a**) (**b**)

Figure 26. High-compact 4 × 4 UWB-MIMO antenna fabricated prototype. (**a**) Top View, (**b**) Bottom View.

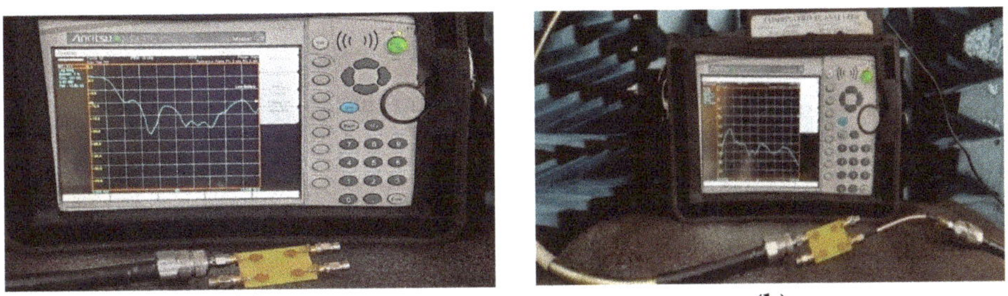

(**a**) (**b**)

Figure 27. High-compact 4 × 4 UWB-MIMO Antenna Measurement. (**a**) S_{11}, (**b**) Isolation.

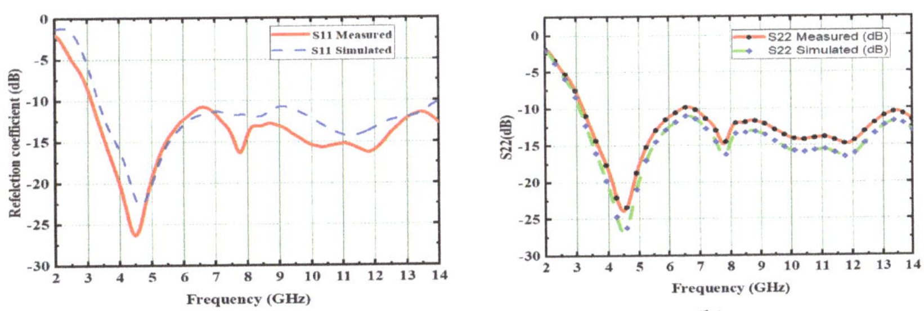

(**a**) (**b**)

Figure 28. Comparison of simulated and measured (**a**) S_{11} and (**b**) S_{22}.

Reduced mutual coupling, high isolation is achieved among all ports; its value is more than 25 dB and it is depicted in Figure 29. The Fields of the 4 × 4 MIMO at 4.5 GHz and 6.8 GHz were measured and are depicted in Figure 30. The E- and H-plane radiation patterns of the radiator are obtained by applying excitation at port one while the remaining ports are matched with a 50Ω load. The obtained radiation patterns are similar for radiator 2, radiator 3, and radiator4 at respective ports 2, 3, and 4. With the help of DRH20, the radiation pattern is measured in an anechoic chamber. The pattern in the H-plane would be almost omni-directional, whereas the pattern in the E-plane is bi-directional.

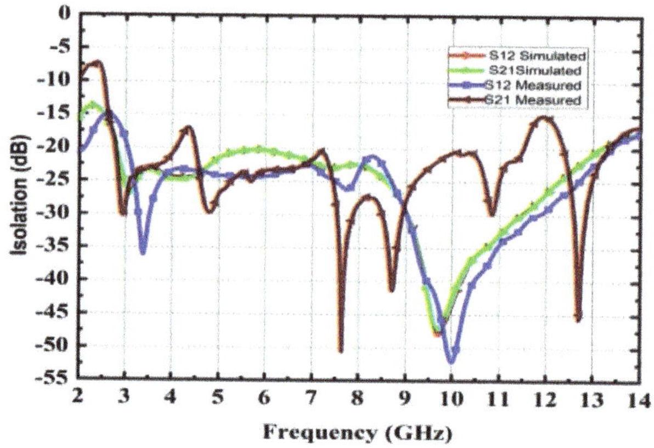

Figure 29. Isolation of 4 × 4 UWM-MIMO antenna.

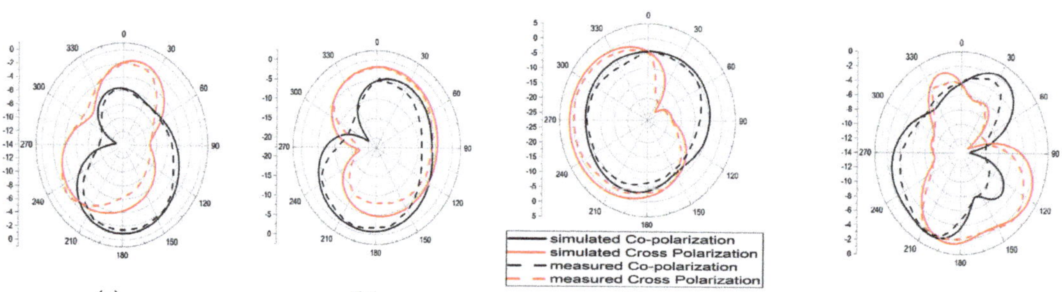

(**a**) (**b**) (**c**) (**d**)

Figure 30. The radiation pattern of E-plane at (**a**) 4.5 GHz and (**b**) 6.8 GHz, and H-plane at (**c**) 4.5 GHz and (**d**) 6.8 GHz.

5.1. 4 × 4 MIMO Gain and Radiation Efficiency

The gain and radiation efficiency of a high-compact 4 × 4 UWB-MIMO antenna is 4.8 dB and 92.2%, respectively. The high-compact proposed 4 × 4 UWB-MIMO antenna radiation, when efficiency simulated and measured, varies from 92.2% to 94% and 90.2% to 92.2%, respectively. These measured values are slightly different from simulated values because of connector losses and fabrication losses in the antenna. The gain and radiation efficiency of the designed high-compact 4 × 4 MIMO antenna is depicted in Figure 31.

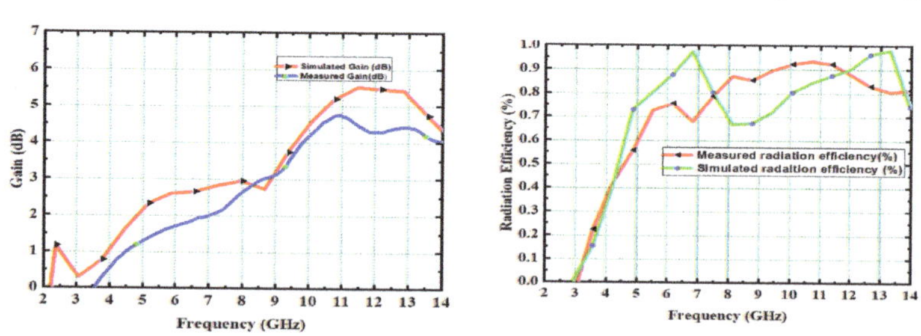

(**a**) (**b**)

Figure 31. High-compact 4 × 4 UWB-MIMO antenna. (**a**) Gain, (**b**) Radiation efficiency.

5.2. Diversity Parameters

The ECC, DG, MEG, and CCL are known as the diversity parameters of the MIMO antenna [37–39]. These parameters should be in the acceptable range only, otherwise the performance of MIMO is not good and is therefore unsuitable for reliable wireless communication. Therefore, the above diversity parameters are calculated and plotted with the help of MATLAB code, and they are also verified with measured values.

5.2.1. ECC and DG

One of the key performances of MIMO antennas is ECC. The acceptable ECC value in MIMO is below 0.5, but by using a systematic design process with the CMA technique, the proposed high-compact 4 × 4 UWB-MIMO antenna achieved a 0.0045 value across the entire UWB spectrum. The ECC explains how the radiating elements interact with each other. No interaction between the radiating elements means that ECC is 0. The ECC is calculated from S-parameters [37].

$$ECC = \frac{\left|S_{ii}^{*} S_{ij} + S_{ji}^{*} S_{jj}\right|^2}{\left(1 - \left(\left|S_{ii}\right|^2 + \left|S_{ji}\right|^2\right)\right)\left(1 - \left(\left|S_{jj}\right|^2 + \left|S_{ij}\right|^2\right)\right)} \quad (4)$$

where S_{ii}, S_{ij}, S_{ji}, and S_{jj} are the S-parameters of UWB-MIMO. This Equation (4) is correct for a 2 × 2 antenna. If MIMO has more than two radiating elements, it is better to calculate ECC with the help of Avg powers along with its directions. The ECC is also represented with the symbol ρ. This represents how the radiating elements are correlated with each other. It measures how the *i*th radiating element is correlated with the *j*th element in the designed MIMO antenna. The method of calculating ECC using S-parameters is not advisable because most of the microstrip antennas used by planners and printers are lossy. Therefore, we can calculate ECC from far-field [38] radiation patterns.

$$\rho_{ij} = \frac{\left|\int_0^{2\pi}\int_0^{\pi}[XPR.E_{\theta i}E_{\theta j}^{*}P_{\theta} + E_{\Phi i}E_{\Phi j}^{*}P_{\Phi}]d\Omega\right|^2}{\int_0^{2\pi}\int_0^{\pi}[[XPR.E_{\theta i}E_{\theta i}^{*} + E_{\Phi i}E_{\Phi i}^{*}P_{\Phi}]d\Omega X \int_0^{2\pi}\int_0^{\pi} XPRE_{\theta j}E_{\theta j}^{*}P_{\theta} + E_{\Phi J}E_{\Phi J}^{*}P_{\Phi}]d\Omega} \quad (5)$$

XPR is cross-polarization rate (XPR = P_V/P_H). The first port is fed with the signal, whereas all remaining ports are terminated and matched with a 50 Ω load. The impact of fading can be reduced by composing the antenna with several radiating elements using a diversity technique. The DG is also an important diversity parameter. The DG is also expressed [37] in terms of ECC.

$$DG = 10\sqrt{1 - ECC^2} \quad (6)$$

The ECC and DG measured and simulated values are plotted and depicted in Figure 32. Here, the ECC is 0.0045, achieved for the proposed antenna; this is a very low value and indicates the low correlation between the MIMO elements. The acceptable value of DG in MIMO is 9.59 dB. The proposed high-compact 4 × 4 UWB-MIMO antenna achieved more than 9.982. The low ECC and high isolation in the MIMO system increase the reliability of the communication system.

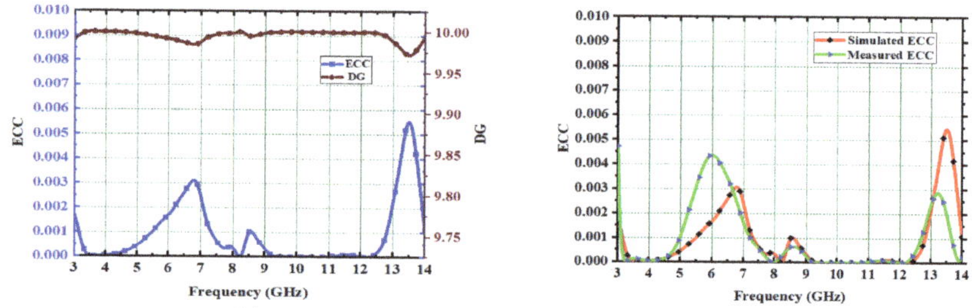

(a) (b)

Figure 32. (**a**) ECC with DG, (**b**) ECC without DG.

5.2.2. MEG

One of the performance matrices of an MIMO antenna is the MEG [38]. The MEG is expressed as:

$$MEG = \int_0^{2\pi} \int_0^{\pi} [\frac{XPR}{1+XPR} F_\alpha(\alpha,\beta) P_\alpha(\alpha,\beta) + \frac{1}{1+XPR} F_\beta(\alpha,\beta) P_{\beta\Phi}(\alpha,\beta)] \sin\alpha d\alpha d\beta \quad (7)$$

where F_α and F_β are patterns of power in the MIMO antenna system, respectively. Generally, in the MIMO antenna system, the MEG between two ports is always less than −3 dB. In the proposed 4 × 4 antenna, the MEG measured between port 1 and port 3 is −3.1 dB, as depicted in Figure 33a.

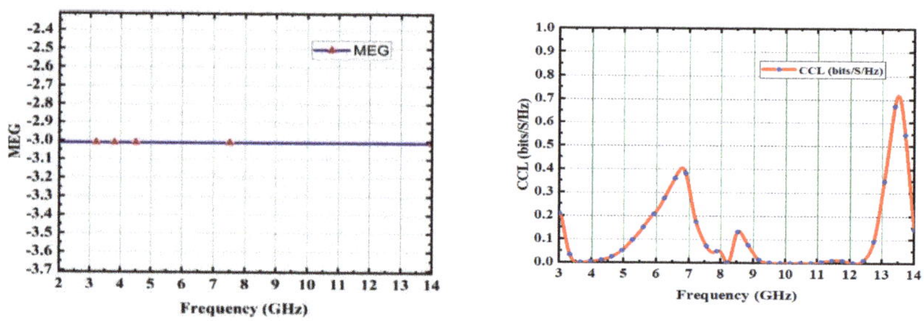

(a) (b)

Figure 33. (**a**) MEG, (**b**) CCL.

5.2.3. The Channel Capacity Loss

The CCL matrices are an important indicator of radiation performance in the MIMO antenna system. The minimum and acceptable limit of CCL in the MIMO system is 0.4 bits/s/Hz. It is mathematically defined as [39]:

$$\text{Closs} = -\log_2 \det(Y^R) \quad (8)$$

$$Y = \begin{bmatrix} Y11 & Y12 & Y13 & Y14 \\ Y21 & Y22 & Y23 & Y24 \\ Y31 & Y32 & Y33 & Y34 \\ Y41 & Y42 & Y43 & Y44 \end{bmatrix} \quad (9)$$

and also expressed as:

$$Y_{nm=-(S^*_{nn}S_{nm}+S^*_{mn}S_{nm})} \quad (10)$$

Here n, m = 1 to 4. The simulated and examined values of CCL are included in Figure 33b. Over the entire UWB frequency spectrum, the designed antenna achieved a CCL of 0.399 bits/s/Hz.

6. Comparisons of Proposed Work with other Models

The performance parameters of the proposed UWB-MIMO antennas are compared with existing UWB-MIMO antenna working models and discussed in the literature, and they are shown in Table 2. The performances of UWB-MIMO antennas were compared, taking into consideration parameters such as the compactness of the antenna, i.e., size and area, design approach, isolation parameters, ECC, channel capacity loss, and MEG, as shown in Table 2. The major advantages of this proposed antenna are good isolation, bandwidth, and the low ECC achieved using CMA.

Table 2. Comparative Analysis with Literature.

Ref.	Design Methodology	Dimensions (mm^2)	Isolation (dB)	ECC	CCL (b/s/Hz)
[2]	Inverted L-shaped strip	60 × 30	34	0.002	0.5
[3]	Decoupling stubs in hexagonal shape	34 × 20	20	0.2	—
[4]	I-shaped decoupling stub	25 × 25	17	0.01	—
[6]	Fractal slot and flipped in horizontal	53 × 35	19	0.007	—
[7]	cut in substrate and placed orthogonal	58 × 28	20	0.008	—
[8]	U-Shaped with T-shaped decoupling stub	30 × 45	29.5	0.00185	—
[19]	Pattern diversity with CMT	85 × 50	20	0.03	—
[23]	CSRR and decoupling stub inverted L-shape	23 × 29	15	0.15	—
[30]	T-shaped Stub	18 × 36 × 18.5	20	0.02	—
[32]	Space diversity	60 × 60	16	0.005	—
[35]	Characteristic Mode Analysis (CMA) with vias and fork-shaped stubs	20 × 28	23	0.005	—
Prop (2 × 2)	CMA with T-shaped decoupling Stub	28 × 20 × 1.6	23	0.005	0.029
[5]	Placement in orthogonal form	38 × 38	20	0.01	0.4
[9]	Polarized diversity with defected ground system	38.3 × 38.3	17	0.02	—
[10]	Fractal slot and flipped in horizontal	40 × 40	12	0.07	0.4
[32]	Space diversity and pattern diversity	90 × 60	20	0.005	—
[35]	Characteristic Mode Analysis (CM) with vias and Fork-shaped stubs	40 × 28	17	0.01	—
[39]	Parasitic decoupler	60 × 60 × 1.52	21	0.001	-
Prop (4 × 4)	CMA with T-shaped decoupling Stub	28 × 40 × 1.6	25	0.0045	0.39

7. Conclusions

The design of compact stub-based UWB-MIMO antennas (2 × 2 and 4 × 4 antennas) are carried out using Characteristic Mode Analysis. Using CMA, a systemic design approach has been achieved. The proposed 2 × 2 and 4 × 4 UWB-MIMO antennas achieve a high isolation of 23 dB and 25 dB, and their bandwidths are 10.11 GHz and 10.775 GHz, respectively. The IBW of compact 2 × 2 and 4 × 4 UWB-MIMO is 124.15% and 126.40%, respectively. The diversity parameters of 2 × 2 and 4 × 4 UWB-MIMO antennas, such as ECC, CCL, MEG, and DG, are 0.005 and 0.0045, 0.39 and 0.39, −3.1 dBi and −3.11 dBi, and 9.978 and 9.982, respectively. The overall performance of both 2 × 2 and 4 × 4 antennas meet the requirements of wireless communication.

Author Contributions: Conceptualization: A.C.S., T.S.R., B.T.P.M. and S.D.; methodology: S.D., T.S.R. and S.L.; software: A.C.S., T.S.R. and A.D.A.; validation: A.C.S., W.E.-S. and S.L.; writing—original draft preparation: A.C.S., B.T.P.M. and S.D.; writing—review, and editing: S.L., A.D.A. and W.E.-S.; supervision: B.T.P.M. and T.S.R.; project administration: W.E.-S.; funding acquisition: A.D.A. All authors have read and agreed to the published version of the manuscript.

Funding: This work is supported by Princess Nourah bint Abdulrahman University Researchers Supporting Project number (PNURSP2022R51), Princess Nourah bint Abdulrahman University, Riyadh, Saudi Arabia.

Data Availability Statement: The data presented in this research are available on request from the corresponding author.

Acknowledgments: The authors would like to acknowledge the Princess Nourah bint Abdulrahman University Researchers Supporting Project number (PNURSP2022R51), Princess Nourah bint Abdulrahman University, Riyadh, Saudi Arabia.

Conflicts of Interest: The authors declare no conflict of interest.

References

1. Awan, W.A.; Zaidi, A.; Hussain, M.; Hussain, N.; Syed, I. The design of a wideband antenna with notchingcharacteristics for small devices using a genetic algorithm. *Mathematics* **2021**, *9*, 2113. [CrossRef]
2. Jayanthi, K.; Kalpana, A.M. Mutual coupling reduction techniques between MIMO antennas for UWB. *Int. J. Recent Innov. Trends Comput. Commun.* **2017**, *5*, 18–22. [CrossRef]
3. Nadeem, I.; Choi, D.Y. Study on Mutual Coupling Reduction Technique for MIMO Antennas. *IEEE Access* **2018**, *7*, 563–586. [CrossRef]
4. Liu, F.; Guo, J.; Zhao, L.; Huang, G.L.; Li, Y.; Yin, Y. Dual-band meta surface-based decoupling method for two closely packed dual-band antennas. *IEEE Trans. Antennas Propag.* **2019**, *68*, 552–557. [CrossRef]
5. Radhi, A.H.; Nilavalan, R.; Wang, Y.; Al-Raweshidy, H.S.; Eltokhy, A.A.; Ab Aziz, N. Mutual coupling reduction with a wideband planar decoupling structure for UWB–MIMO antennas. *Int. J. Microw. Wirel. Technol.* **2018**, *10*, 1143–1154. [CrossRef]
6. Mohanty, A.; Sahu, S. High isolation two-port compact MIMO fractal antenna with Wi-Max and X-band suppression characteristics. *Int. J. RF Microw. Comput.-Aided Eng.* **2019**, *30*, e22021. [CrossRef]
7. Mathur, R.; Dwari, S. A compact UWB-MIMO with dual grounded CRR for isolation improvement. *Int. J. RF Microw. Comput.-Aided Eng.* **2018**, *29*, e21500. [CrossRef]
8. Prabhu, T.; Pandian, S.C. Design and Implementation of T-Shaped Planar Antenna for MIMO Applications. *Comput. Mater. Contin.* **2021**, *69*, 2549–2562. [CrossRef]
9. Bukkawar, S.; Ahmed, V. Compact slot-loaded ultra-wideband multiple-input multiple-output antenna with fractal-inspired isolator. *Int. J. RF Microw. Comput.-Aided Eng.* **2019**, *30*, e22036. [CrossRef]
10. Raheja, D.K.; Kanaujia, B.K.; Kumar, S. Compact four-port MIMO antenna on slottededge substrate with dual-band rejection characteristics. *Int. J. RF Microw. Comput.-Aided Eng.* **2019**, *29*, e21756. [CrossRef]
11. Biswal, S.P.; Das, S. A compact printed ultra-wideband multiple-input multiple-output prototype with band-notch ability for WiMAX, LTEband43, and WLAN system. *Int. J. RF Microw. Comput.-Aided Eng.* **2019**, *10*, e21673. [CrossRef]
12. Gomez-Villanueva, R.; Jardon-Aguilar, H. Compact UWB uniplanar four-port MIMO antenna array with rejecting band. *IEEE Antennas Wirel. Propag. Lett.* **2019**, *18*, 2543–2547. [CrossRef]
13. Premalatha, J.; Sheela, D. Compact four-port vertically polarized UWB monopole antenna for MIMO Communications. *Circuit World* **2021**, *47*, 129–137. [CrossRef]
14. Tiwari, R.N.; Singh, P.; Kanaujia, B.K.; Srivastava, K. Neutralization technique based two and four port high isolation MIMO antennas for UWB communication. *AEU-Int. J. Electron. Commun.* **2019**, *110*, 152828. [CrossRef]
15. Nikam, B.V.; Jadhav, M.R. A Compact Quad Port Band-Notched MIMO Antenna for Wi-Max Applications with Low Mutual Coupling. *Prog. Electromagn. Res. C* **2020**, *104*, 53–67. [CrossRef]
16. Khan, A.A.; Naqvi, S.A.; Khan, M.S.; Ijaz, B. Quad port miniaturized MIMO antenna for UWB 11 GHz and 13 GHz frequency bands. *AEU-Int. J. Electron. Commun.* **2021**, *131*, 153618. [CrossRef]
17. Yang, R.; Xi, S.; Cai, Q.; Chen, Z.; Wang, X.; Liu, G. A Compact Planar Dual-Band Multiple-Input and Multiple-Output Antenna with High Isolation for 5G and 4G Applications. *Micromachines* **2021**, *12*, 544. [CrossRef] [PubMed]
18. Garbacz, R.J. Modal expansions for resonance scattering phenomena. *Proc. IEEE* **1965**, *53*, 856–864. [CrossRef]
19. Barzegari, S.; Forooraghi, K.; Abbasi Arand, B. Design of circularly polarized planar leaky-wave antenna using characteristic mode analysis. *IET Microw. Antennas Propag.* **2021**, *15*, 1086–1099. [CrossRef]
20. Harrington, R.; Mautz, J. Computation of characteristic modes for conducting bodies. *IEEE Trans. Antennas Propag.* **1971**, *9*, 629–639. [CrossRef]
21. Kumar, P.; Ali, T.; MM, M.P. Characteristic Mode Analysis-Based Compact Dual Band-Notched UWB MIMO Antenna Loaded with Neutralization Line. *Micromachines* **2022**, *13*, 1599. [CrossRef] [PubMed]

22. Cabedo-Fabres, M.; Antonino-Daviu, E.; Valero-Nogueira, A.; Bataller, M.F. The theory of characteristic modes revisited. A contribution to the design of antennas for modern applications. *IEEE Antennas Propag. Mag.* **2020**, *5*, 52–68. [CrossRef]
23. Li, K.; Shi, Y. Wideband MIMO handset antenna design based on theory of characteristic modes. *Int. J. RF Microw. Comput.-Aided Eng.* **2018**, *28*, e21217. [CrossRef]
24. Kumar, N.; Khanna, R. A compact multi-band multi-input multi-output antenna for 4G/5G and IoT devices using theory of characteristic modes. *Int. J. RF Microw. Comput.-Aided Eng.* **2020**, *30*, e22012. [CrossRef]
25. Zhao, X.; Yeo, S.P.; Ong, L.C. Planar UWB MIMO antenna with pattern diversity and isolation improvement for mobile platform based on the theory of characteristic modes. *IEEE Trans. Antennas Propag.* **2018**, *66*, 420–425. [CrossRef]
26. Nie, L.Y.; Lin, X.Q.; Yang, Z.Q.; Zhang, J.; Wang, B. Structure-shared planar UWB MIMO antenna with high isolation for mobile platform. *IEEE Trans. Antennas Propag.* **2019**, *67*, 2735–2738. [CrossRef]
27. Medkour, H.; Lakrit, S.; Das, S.; Madhav, B.T.P.; VasuBabu, K. A Compact Printed UWB MIMO Antenna with Electronically Reconfigurable WLAN Band-Notched Characteristics. *J. Circuits Syst. Comput.* **2021**, *31*, 2250045. [CrossRef]
28. Addepalli, T.; Anitha, V.R. A very compact and closely spaced circular shaped UWB MIMO antenna with improved isolation. *AEU-Int. J. Electron. Commun.* **2020**, *114*, 153016. [CrossRef]
29. Alharbi, A.G.; Rafique, U.; Ullah, S.; Khan, S.; Abbas, S.M.; Ali, E.M.; Alibakhshikenari MDalarsson, M. Novel MIMO Antenna System for Ultra-Wideband Applications. *Appl. Sci.* **2022**, *12*, 3684. [CrossRef]
30. Zaidi, A.; Awan, W.A.; Ghaffar, A.; Alzaidi, M.S.; Alsharef, M.; Elkamchouchi, D.H.; Ghoneim, S.S.; Alharbi, T.E. A Low Profile Ultra-Wideband Antenna with Reconfigurable Notch Band Characteristics for Smart Electronic Systems. *Micromachines* **2022**, *13*, 1803. [CrossRef]
31. Awan, W.A.; Hussain, N.; Ghaffar, A.; Zaidi, A.; Naqvi, S.I.; Li, X.J. Compact Flexible Frequency Reconfigurable Antenna for Heterogeneous Applications. In Proceedings of the 2020 9th Asia-Pacific Conference on Antennas and Propagation (APCAP), Xiamen, China, 4–7 August 2020; pp. 1–2. [CrossRef]
32. Khan, M.K.; Khan, M.I.; Ahmad, I.; Saleem, M. Design of a Printed Monopole Antenna with Ridged Ground for Ultra Wide Band Applications. In Proceedings of the 2016 Progress in Electromagnetics Research (PIERS), Shanghai, China, 8–11 August 2016; pp. 4394–4396. [CrossRef]
33. Rahman, S.U.; Cao, Q.; Ullah, H.; Khalil, H. Compact design of trapezoid shape monopole antenna for SWB application. *Microw. Opt. Technol. Lett.* **2019**, *61*, 1931–1937. [CrossRef]
34. Khan, M.S.; Iftikhar, A.; Shubair, R.M.; Capobianco, A.D.; Braaten, B.D.; Anagnostou, D.E. A four element, planar, compact UWB MIMO antenna with WLAN band rejection capabilities. *Microw. Opt. Technol. Lett.* **2020**, *62*, 3124–3131. [CrossRef]
35. Singh, H.V.; Tripathi, S. Compact UWB MIMO antenna with Fork-shaped stub with vias based coupling current steering (VBCCS) to enhance isolation using CMA. *AEU-Int. J. Electron. Commun.* **2021**, *129*, 153550. [CrossRef]
36. Suresh, A.C.; Reddy, T.S. High Isolation With Fork-Shaped Stub In Compact UWB-MIMO Antenna Using CMA. In Proceedings of the 2021 International Conference on Recent Trends on Electronics, Information, Communication & Technology (RTEICT), Bengaluru, India, 27–28 August 2021; pp. 505–511. [CrossRef]
37. Sultan, K.S.; Abdullah, H.H. Planar UWB MIMO-diversity antenna with dual notch characteristics. *Prog. Electromagn. Res. C* **2019**, *93*, 119–129. [CrossRef]
38. Abdulkawi, W.M.; Malik, W.A.; Rehman, S.U.; Aziz, A.; Sheta, A.F.A.; Alkanhal, M.A. Design of a Compact Dual-Band MIMO Antenna System with High-Diversity Gain Performance in Both Frequency Bands. *Micromachines* **2021**, *12*, 383. [CrossRef]
39. Abbas, A.; Hussain, N.; Sufian, A.; Jung, J.; Park, S.M.; Kim, N. Isolation and Gain Improvement of a Rectangular Notch UWB-MIMO Antenna. *Sensors* **2022**, *22*, 1460. [CrossRef]

micromachines

MDPI

Article

Four-Port Dual-Band Multiple-Input Multiple-Output Dielectric Resonator Antenna for Sub-6 GHz 5G Communication Applications

Upesh Patel and Trushit Upadhyaya *

Electronics and Communication Department, Chandubhai S. Patel Institute of Technology, Charotar University of Science and Technology (CHARUSAT), Changa 388421, India
* Correspondence: trushitupadhyaya.ec@charusat.ac.in

Abstract: A four-port dielectric resonator (DR)-based multiple-input multiple-output (MIMO) antenna is presented for sub-6 GHz MIMO communication. The dielectric resonator was fed through aperture feeding to achieve dual-band resonance. The DRA has the operating modes of $TE_{01\delta}$ and $TE_{10\delta}$ at 3.3 GHz and 3.9 GHz, respectively. The engineered antenna has port isolation of higher than 20 dB at the target frequencies without the employment of an extra isolation mechanism. Full-wave high-frequency simulation software was employed for the simulation computation. The antenna has a peak gain of 5.8 dBi and 6.2 dBi, and an efficiency of 88.6% and 90.2% at 3.3 GHz and 3.9 GHz, respectively. The proposed resonator has good MIMO diversity parameters. The optimal envelope correlation coefficient (ECC) is 0.01, channel capacity loss (CCL) is 0.1 bits/sec/Hz, and the total active reflection coefficient (TARC) is −22.46. The DRA elements are aligned orthogonally with adequate displacement for achieving polarization diversity and spatial diversity. The antenna delivers its applications in Sub-6 GHz 5G and WiMAX communications.

Keywords: dielectric resonator; MIMO communication; 5G antenna; Sub-6 GHz communication; WiMAX communication

Citation: Patel, U.; Upadhyaya, T. Four-Port Dual-Band Multiple-Input Multiple-Output Dielectric Resonator Antenna for Sub-6 GHz 5G Communication Applications. *Micromachines* **2022**, *13*, 2022. https://doi.org/10.3390/mi13112022

Academic Editor: Mark L. Adams

Received: 4 November 2022
Accepted: 18 November 2022
Published: 19 November 2022

Publisher's Note: MDPI stays neutral with regard to jurisdictional claims in published maps and institutional affiliations.

Copyright: © 2022 by the authors. Licensee MDPI, Basel, Switzerland. This article is an open access article distributed under the terms and conditions of the Creative Commons Attribution (CC BY) license (https://creativecommons.org/licenses/by/4.0/).

1. Introduction

The significantly rising requirement for real-time internet traffic has triggered the requirement for fast and efficient communication. The present data streaming need has to overcome the restrictions offered by 3G and 4G communications. While there is a strict constraint on channel allocation numbers, there also exist severe limitations in multi-user accommodation due to co-channel interference. The substantial evolution in wireless communication has triggered 5G communication having the capability to support data rates up to multiple gigabits per second. Each advanced generation has presented a multifold growth in communication technology and a higher data rate [1,2]; low latency and delay are offered with 5G communication in comparison with 4G communication. The multiple-input multiple-output (MIMO) technology provides increased channel capacity, mitigates multipath fading, and provides faster communication. MIMO provides seamless communication through optimum spectrum efficiency, and through effective channel allocation, it also provides more mobile users.

The 5G communication technology inducts an array of antennas. A MIMO communication tames radio propagation issues by communicating the data over manifold transmit and multiple receive antennas. The multiple fading inconsistency affects the data transmitted over each antenna and hence the fading received in each channel differs in nature. At present, there is a wide spectrum of multiplexing techniques available for giving a high degree of freedom in MIMO communication. The 5G mobile communication has a strong constraint of physical space for incorporating the antennas. Planar resonators are a feasible choice for the sustenance of multiple antenna systems. Patch resonators provide appropriateness in manufacturing and cost-competitiveness. The patch antennas suffer significantly in terms of providing good antenna parameters, especially bandwidth and gain. The standard patch antenna employs high dielectric loss material to provide cost-competitiveness and hence there has to be a balance in the trade-off between antenna parameters and cost-effectiveness. The dielectric resonator antennas (DRAs) are capable of providing enhanced radiation efficiency when they are fed appropriately. The DRA design with proper electrical and mechanical attributes leads to greater radiation efficiency. The literature presents a few of the DRAs designed for 5G MIMO communication [3–10].

MIMO antennas are typically expected to give low mutual coupling between ports. The high isolation or low correlation ensures better radiation and diversity performance in MIMO systems. These antenna properties are expected in a space-constrained environment and it is extremely important to have electrical compactness in the antenna design. This makes the MIMO antenna design exciting. There is significant literature available relating to the improvement of MIMO antenna port isolation through the decoupling mechanism [11–20]. In the absence of the decoupling mechanism, port isolation can be achieved by the appropriate placement of the resonators which are fed through transmission lines. Internal mode decoupling is frequently used for achieving better port isolation. The engineered resonators will guarantee opposite current flow directions in the resonator to avoid a strong correlation between the resonating elements. The WiMAX communication standard provides broadband services to end-users. It is capable of providing a higher data rate compared to WiFi communication over a longer distance. MIMO communication requires multiple resonant frequencies and there has been substantial research in MIMO technology [21–27]. The recent literature presents state-of-the-art MIMO antennas [28–32]. The effective use of a 3–300 GHz underutilized spectrum may aid in providing a reasonable trade-off between user coverage and channel capacity, as presented by 3G and 4G communication. By engineering the dielectric resonator, improved antenna parameters viz. directivity, gain, bandwidth, and cross-pol isolation can be achieved. The adequate selection of antenna dimensions and appropriate feeding mechanisms can ensure the excitation of multiple resonant modes. The novelty of the presented design is that by maneuvering the dielectric resonator two modes viz. $TE_{01\delta}$ and $TE_{10\delta}$ are excited in antenna dual-band operation. The antenna also provides electrical compactness for integration in the RF circuit design. The presented antenna has an excellent application in Sub-6GHz and WiMAX Communication. Section 2 describes antenna design, the antenna sensitivity analysis is presented in Section 3 to show the effects of physical parameters on antenna resonance, and Sections 4 and 5 of the manuscript incorporate implementation on 4 $\times$ 4 MIMO design followed by a critical discussion of the results.

2. Antenna Design

Figure 1 exhibits a designed MIMO antenna. The patch resonator is excited through the transmission line using the co-axial cable. The dielectric resonator (DR) was kept on the patch resonator. The DR is fed by an aperture coupling mechanism. One of the important features of DR is a lack of metal which has huge conducting losses for the GHz frequency spectrum. Customary conductor-based resonators notably suffer from these metallic losses. The DR-based antennas offer extremely high efficiency and hence the high antenna gain. If an appropriate DR material is selected, then DRA provides large bandwidth and gain. Full-wave high-frequency simulation software (HFSS) was employed for the simulation computation. The presented antenna employs alumina material as the dielectric resonator, with ε_r = 9.9 and tanδ = 0.0001. This material is widely available and it provides an economical solution compared to other expensive dielectric materials. The aperture-fed DRA was initially kept on the patch resonator. Several repetitions in simulations were taken up to improve the electrical dimensions of the dielectric material to excite the two modes. The fundamental $TE_{01\delta}$ and $TE_{10\delta}$ mode excitations present the key feature to achieving targeted frequency characteristics. Table 1 presents the mechanical sizes of the dielectric resonator.

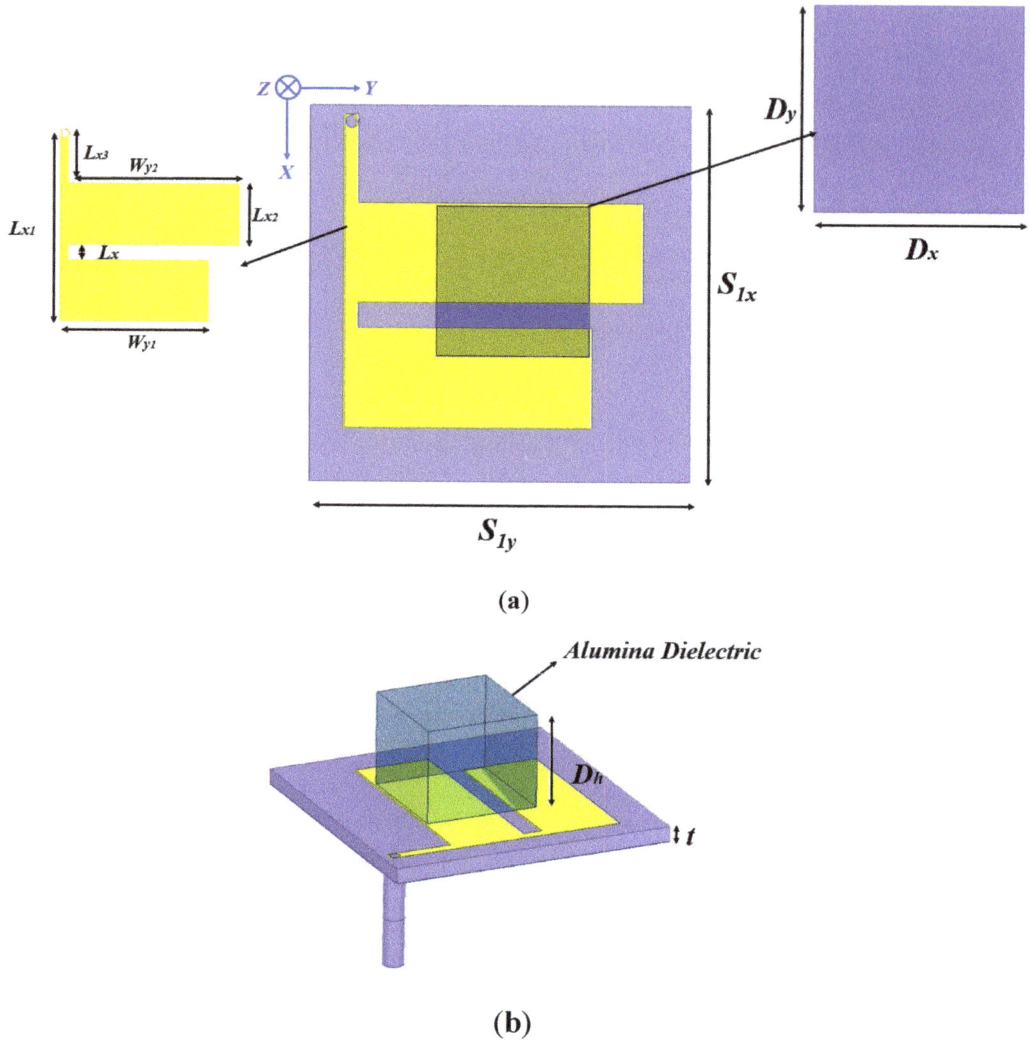

Figure 1. Antenna views, (**a**) top, and (**b**) perspective.

Table 1. Antenna Mechanical Dimensions.

Parameter	Dimensions (mm)	Parameter	Dimensions (mm)
$S_{1x} = S_{1y}$	30	$Lx3$	7
$Dx = Dy$	12	$Wy1$	19.5
Lx	2	$Wy2$	22.4
$Lx1$	25	Dh	10
$Lx2$	8	t	1.6

The electromagnetic fields emanating from the conducting patch are coupled to a dielectric resonator which causes the excitation of the DR. The offset feeding technique is employed for impedance matching. The microstrip feedline dimensions were initially numerically computed and then further optimized for the feeding of the patch antenna. The antenna has been mounted on a standard low-cost FR-4 substrate having a dielectric constant and loss tangent of 4.4 and 0.02, respectively. The 35 μm copper sheets were kept for the design of the patches. The square DR has lateral dimensions of 12 mm and a height of 10 mm. There are few DR-based MIMO antennas in the literature that have more height to attain improved antenna parameters, however, the space constraint may restrict the maximum height dimensions of DRAs. One of the essential reasons to induct a square dielectric resonator instead of a circular/cylindrical dielectric resonator is the mechanical viability in antenna design engineering. The placement optimization of a square DR on the top of the patch is relatively simpler than on cylindrical DRs. It was also observed that it is easier to predict the antenna behavior during the optimization for the case of a square antenna. It is extremely difficult to achieve an antenna numerical model or prediction model for a combination of the patch and dielectric material. It is extremely viable to attain the target antenna parameters through software simulations, however, DRA excitation modes equations can aid in the crude estimation of the antenna resonance. Another reason to avoid circular/cylindrical dielectric resonators is to evade degenerated modes being induced.

Figure 2 exhibits the antenna evolution. For 'Phase-1', the simulated reflection coefficients confirm the impedance bandwidth from 3.3 to GHz and 3.6 to GHz, not covering the desired frequency range. To enhance the bandwidth further and attain the desired resonance, an inverted F-shaped radiator was made by splitting the rectangle into two halves, which is depicted as 'Phase-2'. This increased the electrical length and thus shifts the cut-off frequency to the upper spectrum and improves the impedance bandwidth. Furthermore, to achieve the desired Sub-6 5G and WiMAX band, an inverted F-shaped antenna is further modified by etching the edge of the first rectangle, as depicted in 'Phase-3'. The antenna design steps are as follows: (i) implementation of engineered patch resonator, (ii) secondary mode excitation by inducting the dielectric resonator, (iii) altering the antenna resonance by modifying the electrical dimensions of the resonator through slit gap, and (iv) design optimization through sensitivity analysis.

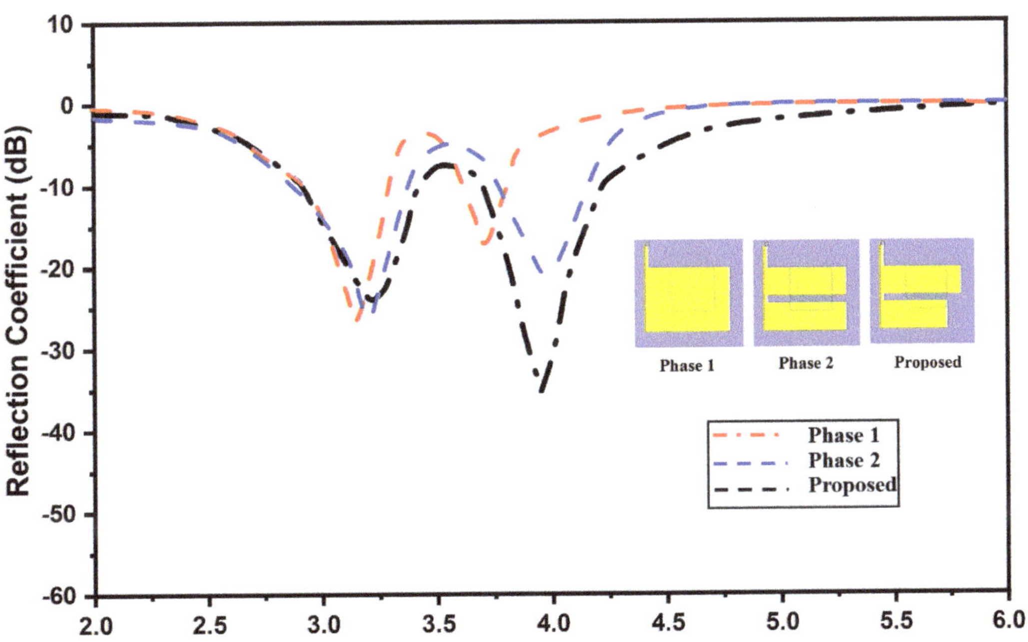

Figure 2. Antenna development phases.

The calculated sizes of the dielectric resonators were estimated through computation using dielectric resonator equations for the generation of $TE_{01\delta}$ and $TE_{10\delta}$ modes to achieve the dual-band performance of the resonator. The resonances at 3.3 GHz and 3.9 GHz were attained. Figure 3 illustrates the electric field generated in the DR through the simulation. There is a dissimilarity in the simulated and numerically computed resonance mode as the dielectric wave model does not keep account of the employed feeding mechanism [8].

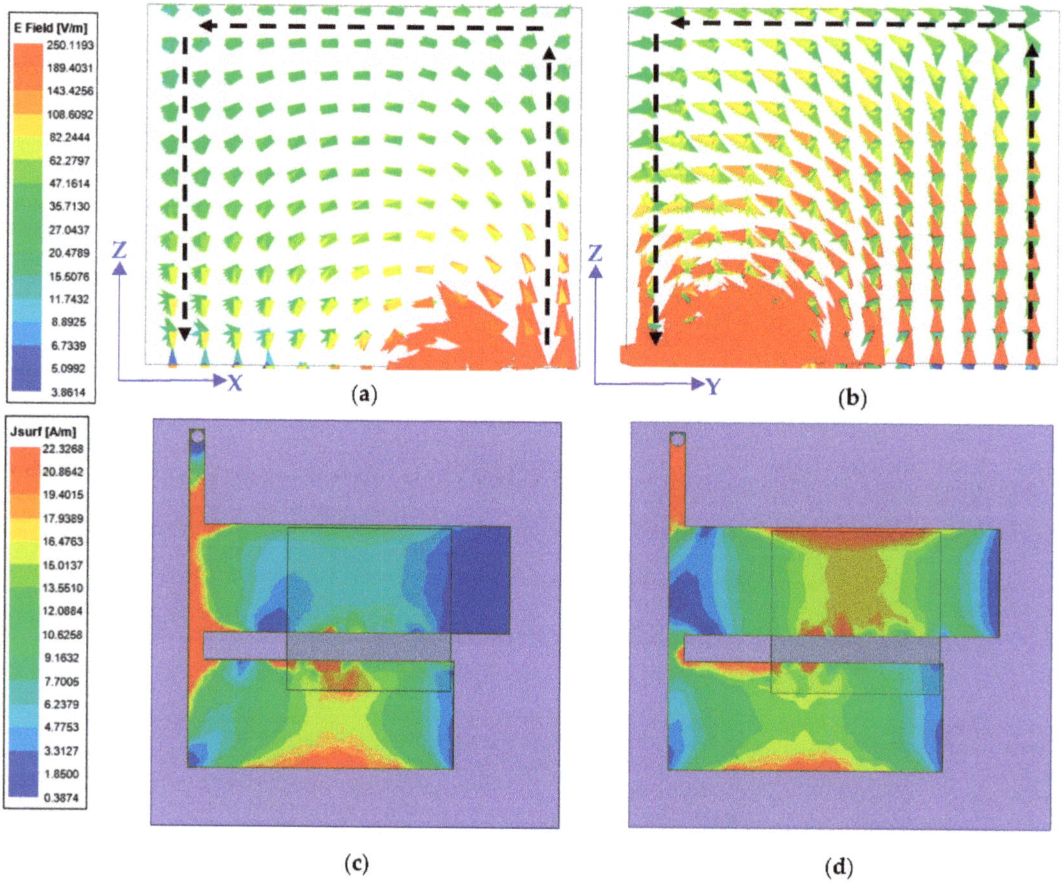

Figure 3. E-Field within DR; (**a**) $TE_{01\delta}$ at 3.3 GHz; (**b**) $TE_{10\delta}$ at 3.96 GHz; (**c**) surface current on resonator at 3.3 GHz; and (**d**) surface current at 3.96 GHz.

Through the software simulations, it was possible to visualize the higher-order modes being produced within the DR. The E-field distribution in the x-plane and y-plane of DR at 3.3 GHz and 3.96 GHz are illustrated in Figure 3a,b. Through these figures, the excitation of the $TE_{01\delta}$ and $TE_{10\delta}$ modes are confirmed. Figure 3c,d depict the surface current density on the resonator.

The resonance and DR mode computations can be done by using the dielectric wave model [8]:

$$k_z tan\left(\frac{k_z d}{2}\right) = \sqrt{(\varepsilon_r - 1)(k_o{}^2) - k_z{}^2} \tag{1}$$

$$k_x{}^2 + k_y{}^2 + k_z{}^2 = \varepsilon_r k_o{}^2 \tag{2}$$

$$k_x = \frac{m\pi}{a}\ ;\ k_y = \frac{n\pi}{b};\ k_o = \frac{2\pi}{\lambda_o} \tag{3}$$

k_o : free-space wavenumber
k_x, k_y, and k_z: half-wave variations
a, b, and d: mechanical dimensions.

3. Design Optimization

Numerous design parameters of the resonator can alter the antenna resonance and associated radiation performance. The prime parameters include the electrical dimensions of the resonators. Intuitively, modifying the total electrical length alters the resonant modes being excited. A range of techniques to improve antenna radiation parameters by altering the antenna dimensions have been presented in the literature.

Numerous simulation iterations were performed to achieve the target antenna parameters. Each physical dimension of the antenna was varied to attain improved antenna radiation parameters. Selected parameter variations are displayed in Figure 4. The dielectric resonator dimensions are vital in attaining the desired resonances. The variations in length and width of the resonators are illustrated in Figure 4a,b. Vital changes in resonances occurred while changing the length and width of the DR. It was noticed that by varying the length and width while keeping the other parameters constant, the net electrical current path of the proposed antenna was greatly affected; thereby, the resonant frequencies were shifted. The parametric variation in resonator height is depicted in Figure 4c. It was perceived that reducing the height of the resonator caused the resonance to increase for either band. An additional variation in the gap between the two flares of the antenna is exhibited in Figure 4d. This variation was significant in the change in resonance frequency apart from the variation in dielectric resonator dimensions. It is quite clear from the plot that the resonating frequencies significantly change by altering the flare distance. The impedance matching radically improves with the value of 2mm. The field coupled from the antenna to the dielectric resonator notably varies with a change in gap size which causes alterations in the antenna resonance. Figure 4e,f provide the resonance shift with a change in the length of the upper and lower flares, respectively. It is evident that due to the change in the effective electrical length of the antenna, there is a major shift in the antenna resonance. While the reflection coefficient and bandwidth increase for one of the bands, the other band does not fall on the target frequencies depending on which optimum dimensions are kept to balance the trade-off. A minor variation in reflection coefficient and resonance frequencies was observed for the change in the flare width, as apparent from Figure 4g,h,e. The length variation in Lx3 does not have any effect on the antenna resonance since the impedance does not vary with a minor change in the line length. As evident from Figure 4j, significant modifications were observed with the change in the location of the offset feed due to the modification to the impedance matching of the antenna at the target resonance.

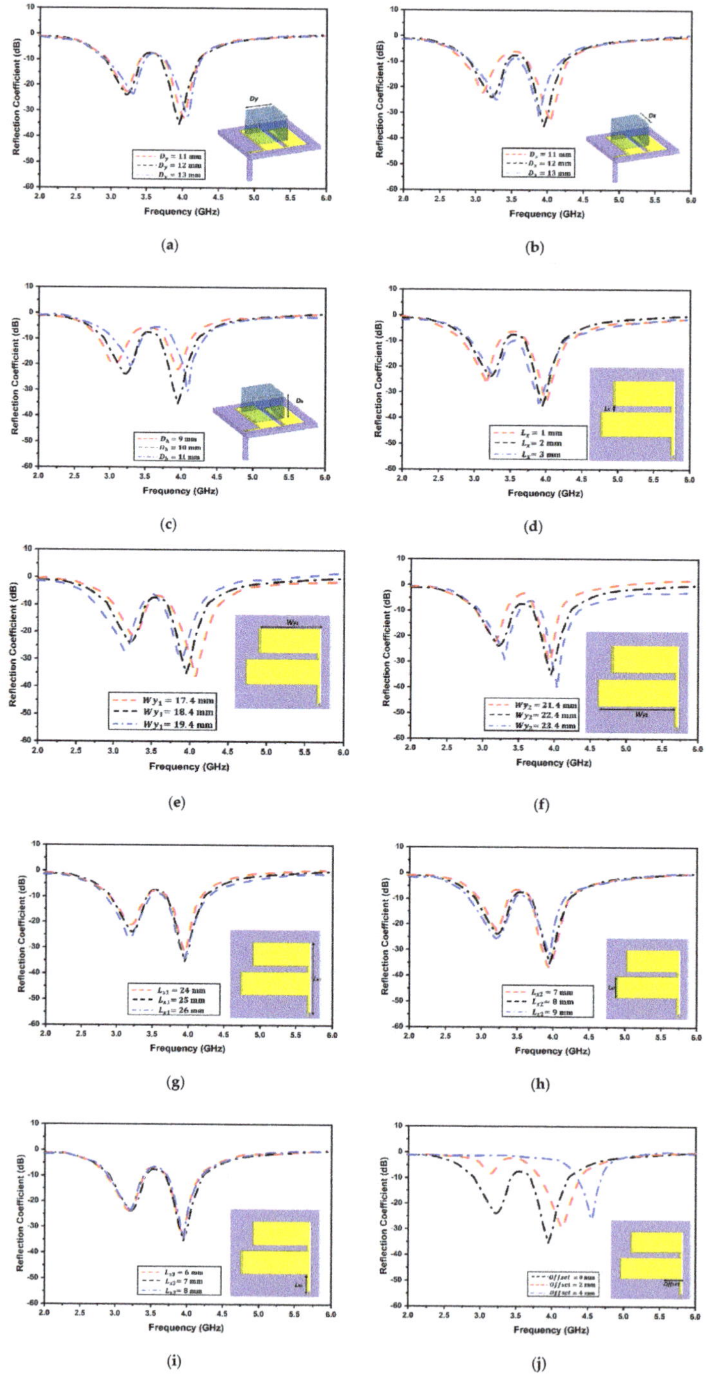

Figure 4. Sensitivity analysis of antenna; (**a**) DR Length D_y; (**b**) DR Width D_x; (**c**) DR Height D_h; (**d**) flare gap; (**e**) upper flare W_{y1}; (**f**) lower flare W_{y2}; (**g**) length L_{x1}; (**h**) width L_{x2}; (**i**) length L_{x3}; (**j**) offset feed location.

4. MIMO Antenna Design Configuration

The proposed MIMO antenna is displayed in Figure 5. The designed MIMO antenna consists of the quad resonating elements of both the dielectric and patch resonators. There needs to be a low correlation between the antenna ports and the low mutual coupling among the four elements for achieving an appropriate diversity performance. Hence, the substrate length and width were modified to the required proportions, keeping the overall electrical dimensions in perspective. The top of the substrate has four identical modified inverted F-shaped radiators placed orthogonally about each other, over which four different alumina dielectric resonators (DR1, DR2, DR3, and DR4) were integrated, whereas, at the back, the full ground profile of the resonator was maintained. Numerous iterative simulations were carried out to achieve electrical compactness without a major trade-off in port isolation. The ground plane and substrate length and width were retained at 60 mm. The overall resonator dimensions are 60 × 60 × 11.6 mm^3. The dielectric resonators are connected to the patch resonator using conducting adhesive.

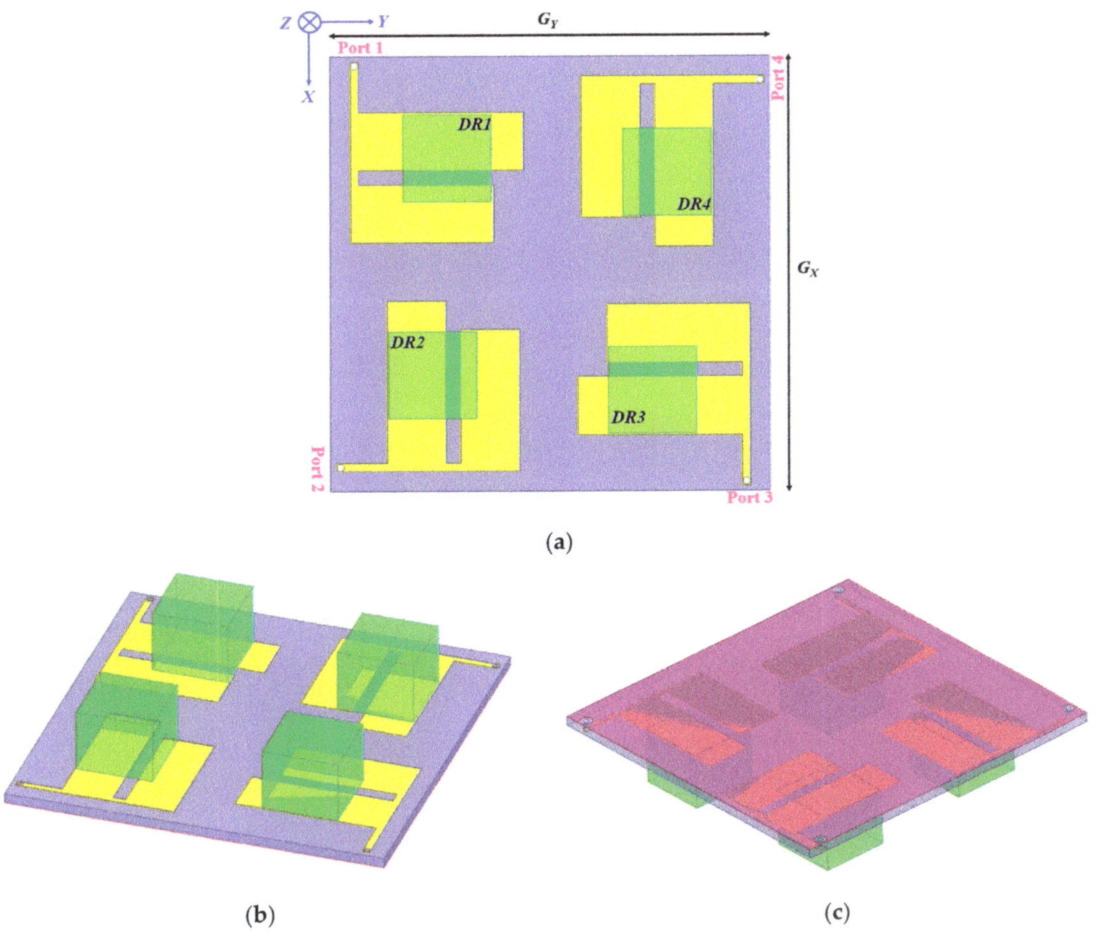

Figure 5. MIMO antenna view; (**a**) top; (**b**) perspective-top; and (**c**) perspective ground.

5. Proposed Antenna Results and Discussions

Figure 6 exhibits the fabricated prototype. The simulated and measured reflection coefficients and mutual coupling plots of the designed Sub-6 5G and WiMAX quad-element

MIMO antenna are demonstrated in Figure 7a,b, respectively. A 2:1 VSWR bandwidths of 620 MHz (2.86 GHz–3.48 GHz) and 580 MHz (3.67 GHz–4.25 GHz) are available over two bands of frequencies. However, mutual coupling among ports has been greatly reduced in the presence of DRA compared to without DRA. The proposed structure covers a Sub-6 GHz 5G and WiMAX frequency spectrum with good inter-element isolation.

(a)

(b)

Figure 6. Fabricated DRA Prototype. (**a**) Top View; (**b**) Back View.

A Keysight E5063A vector network analyzer (VNA) was used to measure the S-parameters of the proposed antenna. These measured results were not the same as the simulated results. The difference in the results was mainly due to manufacturing errors or testing intolerances. From Figure 7, it is also observed that the measured inter-element isolation, i.e., adjacent and diagonal elements, are nearly identical to the simulated isolation results because of the symmetrical structures. The measured isolation is more than −20 dB at 3.3 GHz and less than −26 dB at 3.9 GHz. Furthermore, to validate the proposed antenna's radiation performance, the simulated 2D radiation patterns were compared with the measured 2D antenna patterns using an anechoic chamber. Figure 8 illustrates the 2D far-field radiation patterns of the proposed antenna at the resonating frequencies of 3.3 GHz and 3.9 GHz, respectively. The 2D radiation patterns were calculated in two planes, both E and H. For measuring the far-field radiation patterns, a single radiating element was fed, whereas all others were matched and terminated with an impedance load of 50 Ω. It is noticed from the plot that because of the symmetry of the design, the patterns are mirror images and are orthogonal to each other. Therefore, the proposed design has great potential in offering a good diversity of characteristics.

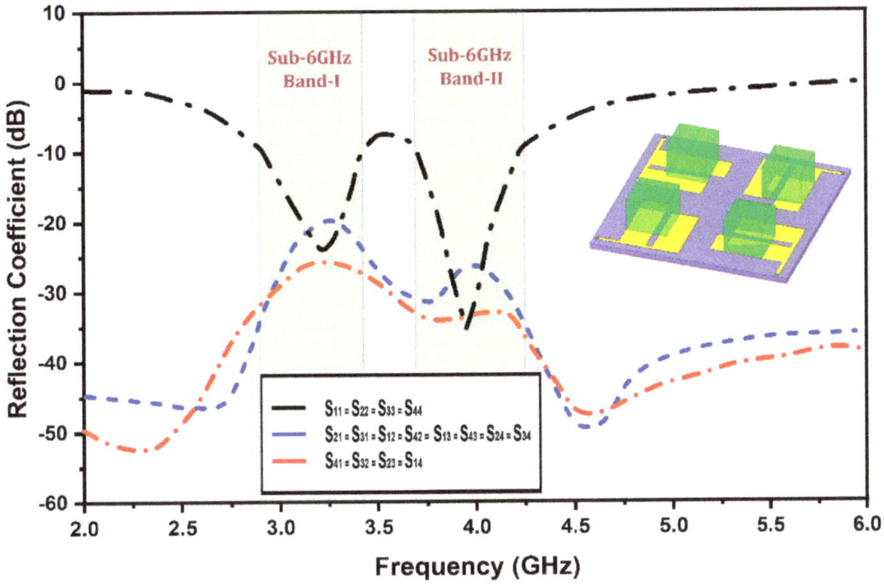

(**a**)

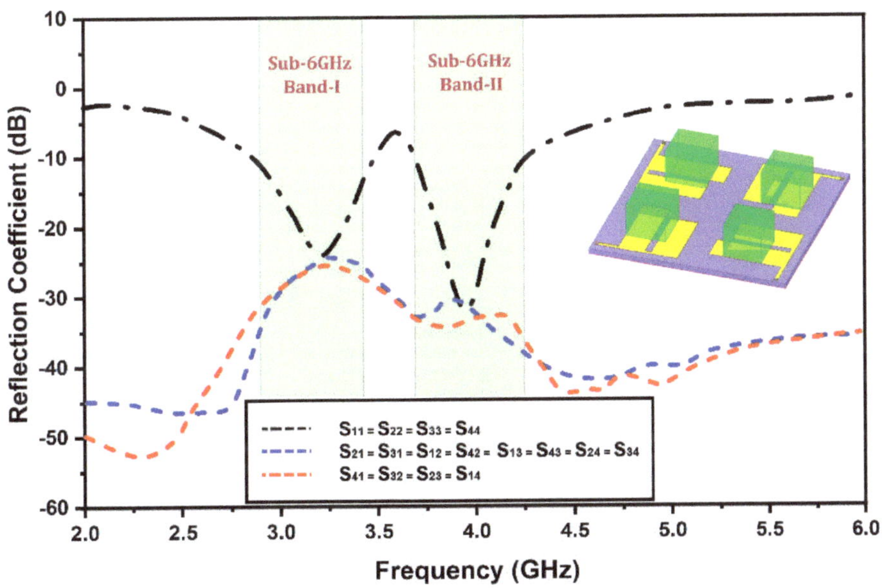

(**b**)

Figure 7. Antenna reflection coefficient; (**a**) simulated; and (**b**) measured.

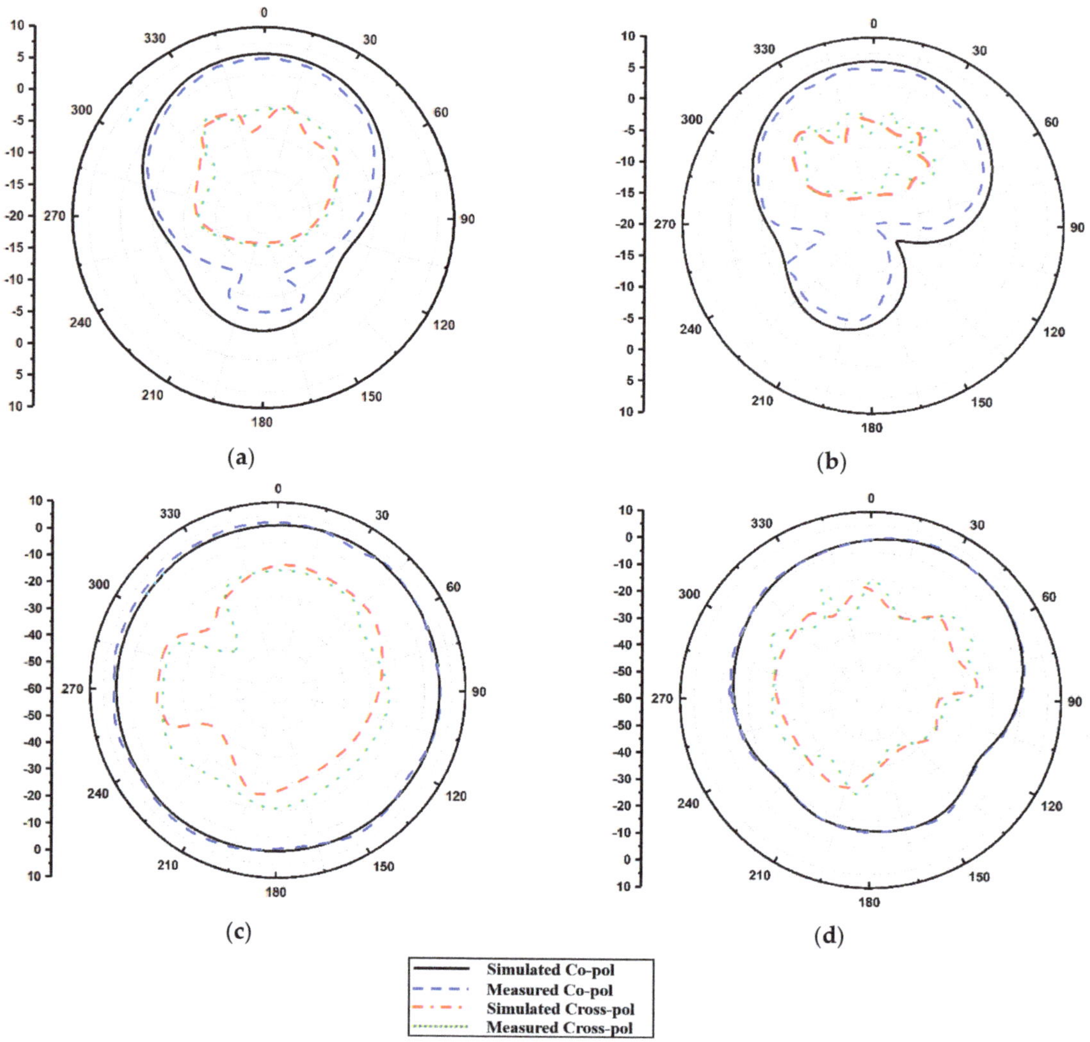

Figure 8. Antenna radiation pattern: (**a**) E-Plane 3.3 GHz, (**b**) E-Plane 3.9 GHz, (**c**) H-Plane 3.3 GHz, and (**d**) 3.9 GHz.

Furthermore, Figure 9 highlights the antenna gain. The simulated gains over the operating frequency bands are equal. However, the measured peak gains over the Sub-6 and WiMAX bands are 5.8 dBi and 6.2 dBi, respectively, slightly different from the simulated antenna gains due to fabrication errors. The antenna efficiency for all four radiators is also exhibited in Figure 9. The plot reveals that the peak efficiencies of the proposed antenna over the Sub-6 and WiMAX bands are 88.6% and 90.2%, respectively.

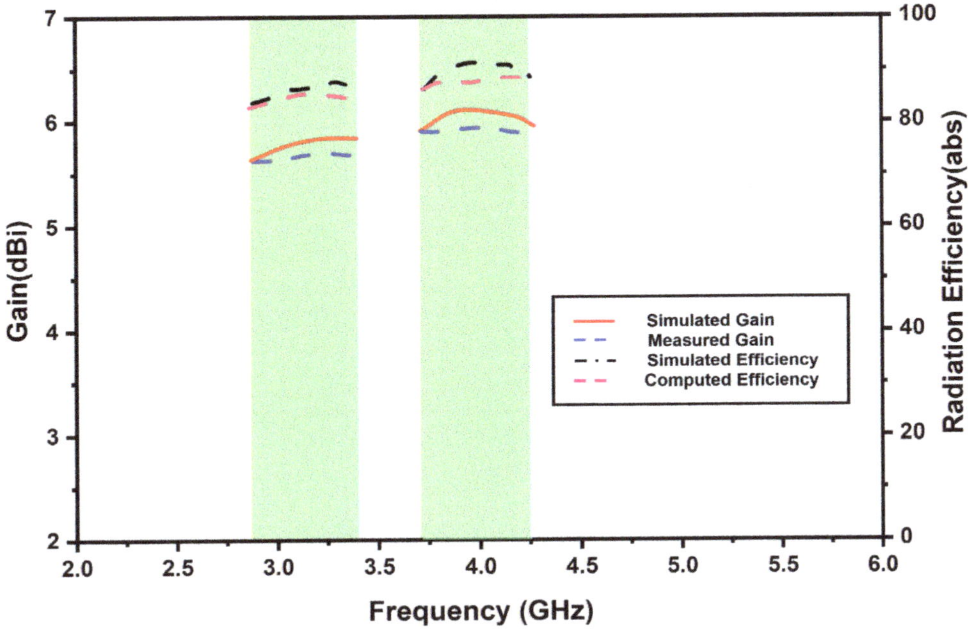

Figure 9. Antenna gain and efficiency against frequency.

The diversity of the proposed antenna systems for Sub-6 5G and WiMAX applications was computed. Various parameters such as channel capacity loss (CCL), mean effective gain (MEG), and envelop correlation coefficients (ECC) of the radiating elements were analyzed. Firstly, ECC was observed, in which the degree of correlation among antenna elements was measured. ECC should not be more than 0.5. The formula for ECC calculation with far-field is expressed in the following equations [8]:

$$\rho_e(i,j) = \frac{\int_0^{2\pi}\int_0^{\pi} A_{i,j}(\theta,\phi)\sin\theta\, d\theta d\phi}{\sqrt{\int_0^{2\pi}\int_0^{\pi} A_{i,i}(\theta,\phi)\sin\theta\, d\theta d\phi} + \sqrt{\int_0^{2\pi}\int_0^{\pi} A_{j,j}(\theta,\phi)\sin\theta\, d\theta d\phi}} \tag{4}$$

$$A_{i,j}(\theta,\phi) = XPRE_{\theta i}(\theta,\phi)E^{*}{}_{\theta j}(\theta,\phi)P_{\theta}(\theta,\phi) + E_{\phi i}(\theta,\phi)E^{*}{}_{\phi j}(\theta,\phi)P_{\phi}(\theta,\phi) \tag{5}$$

$$\rho_{eij} = \frac{\left|S_{ii}^{*} * S_{ij} + S_{ji}^{*} * S_{jj}\right|^2}{(1 - |S_{jj}|^2 - |S_{ji}|^2)(1 - |S_{jj}|^2 - |S_{ij}|^2)} \tag{6}$$

where functions $Ai(\theta, \phi)$ and $Aj(\theta, \phi)$ represent the 3D far-field pattern value obtained when port i and j are excited, respectively. Hermitian product operator is denoted using *, and ω refers to the solid angle. XPR is the cross-polarization level defined as the ratio of average power along the phi and theta directions. Figure 10 depicts the simulated and measured ECC plot (for all ports) of the proposed antenna. Both measured and simulated ECC values over Sub-6 5G and WiMAX bands are less than 0.04 and 0.02, respectively. Such low values of ECC for the proposed antenna ensure a high-diversity gain.

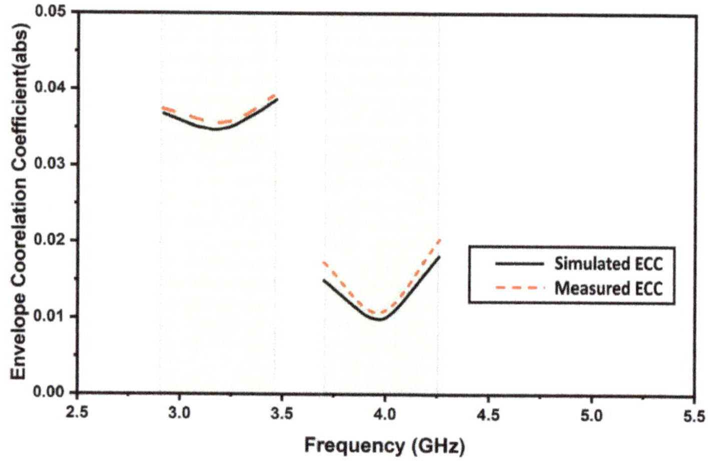

Figure 10. Simulated and measured ECC.

Moreover, to properly characterize the MIMO system, the total active reflection coefficient (TARC) and channel capacity loss (CCL) were analyzed. TARC provides the reflected–incident power ratio, whereas CCL quantifies the upper threshold amount of data rate where the information signal is continuously transmitting without sustaining any notable error. The optimal value of TARC is –22.46; the TARC is shown in Figure 11. The TARC can be calculated as [8]:

$$TARC = \frac{\sqrt{\sum_{n=1}^{N} |b_n|^2}}{\sqrt{\sum_{n=1}^{N} |a_n|^2}} \tag{7}$$

$$b_n = [S]a_n \tag{8}$$

where b_i represents the reflected signal, a_i represents the incident signal, N depicts the number of elements in the MIMO system, and S denotes the scattering parameter.

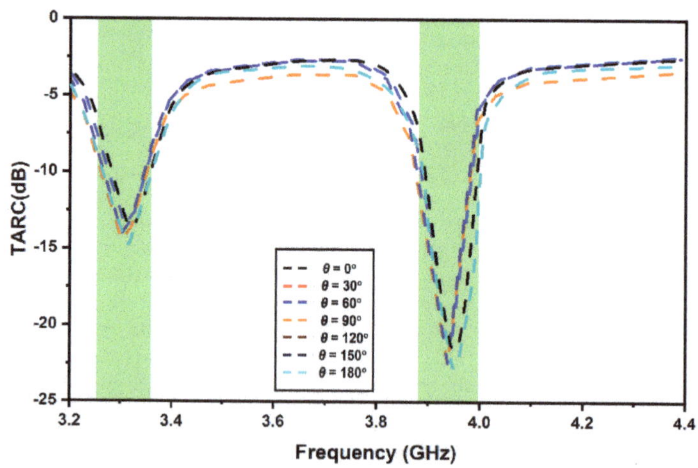

Figure 11. Proposed MIMO antenna TARC.

Thus, ideally, channel capacity loss (CCL) must be zero. Figure 12 depicts CCL. The following expression is used to compute CCL [8]:

$$C_{loss} = -log_2\left|\psi^R\right| \quad (9)$$

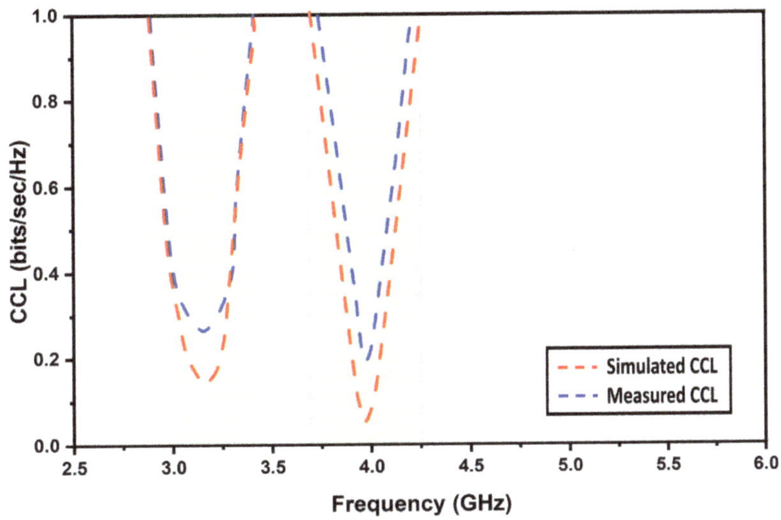

Figure 12. Simulated and measured CCL.

The Ψ^R represents the correlation matrix between the antenna elements at the receiver. It is described as follows in Equation [8]:

$$\Psi^R = \begin{bmatrix} \rho_{11} & \rho_{12} & \rho_{13} & \rho_{14} \\ \rho_{21} & \rho_{22} & \rho_{23} & \rho_{24} \\ \rho_{31} & \rho_{32} & \rho_{33} & \rho_{34} \\ \rho_{41} & \rho_{42} & \rho_{43} & \rho_{44} \end{bmatrix} \quad (10)$$

$$\rho_{ii} = 1 - \sum_{n=1}^{4}\left(S_{in}{}^{*}S_{ni}\right); \rho_{ij} = -\sum_{n=1}^{4}\left(S_{in}{}^{*}S_{nj}\right) \text{ For} i,j = 1,2,3,\text{or } 4. \quad (11)$$

The desired level of CCL should not be more than 0.4 bits/sec/Hz. It is visible from the plot that the CCL values are lower than the expected values over both the Sub-6 5G and WiMAX bands of frequencies, which confirms that the proposed MIMO antenna offers stable diversity characteristics.

Another important diversity parameter is MEG, which measures the average received signal strength of each element. The mathematical expression for MEG calculation is given as [8]:

$$MEG = \int_0^{2\pi}\int_0^{\pi}\left(\frac{XPR}{1+XPR}G_\theta(\theta,\phi)P_\theta(\theta,\phi) + \frac{1}{1+XPR}G_\phi(\theta,\phi)P_\phi(\theta,\phi)\right)\sin\theta \, d\theta d\phi \quad (12)$$

$$MEG_i = 0.5\left[1 - \sum_{j=1}^{N}\left|S_{ij}\right|^2\right] \quad (13)$$

XPR is the cross-polarization level defined as the ratio of average power along the phi and theta directions, where i ranges from 1 to 4, since the proposed antenna has four ports, and $|S_{ij}|$ depicts the scattering parameters of the proposed SIW-fed DRA MIMO antenna. Figure 13 presents the MEGs for all ports of the proposed antenna over Sub-6 5G and WiMAX bands. Here MEG12 (between Port 1 and Port 2) and MEG14 (between Port 1 and Port 4) represent the MEG of the adjacent elements, whereas MEG13 (between Port 1 and Port 3) represents the diagonal elements. It is visible from the plot that the MEGs ratio is almost equal and close to unity. Hence, the proposed antenna fulfills the equality condition of the diverse systems and consequently exhibits excellent MEG characteristics, since the deviations among the MEGs are well under 3dB. Table 2 exhibits the antenna comparison with the selected literature, and evidently, the presented antenna demonstrates better characteristics.

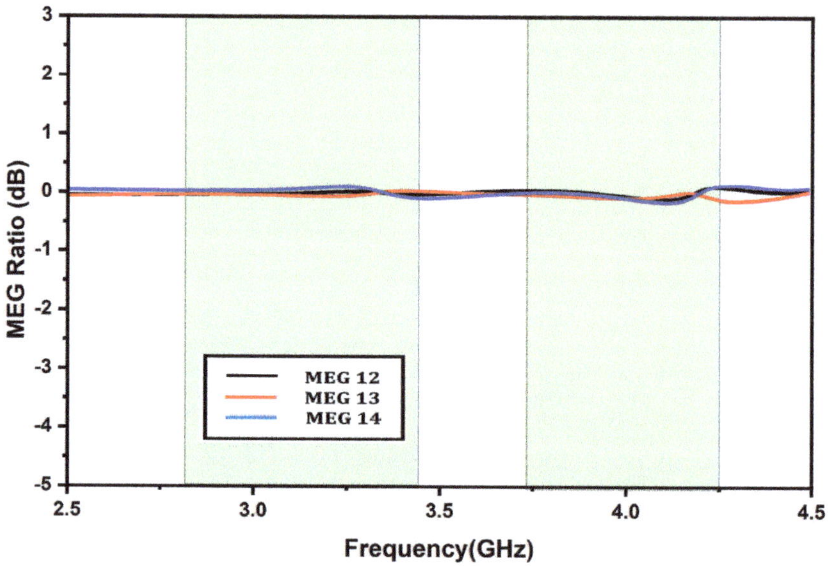

Figure 13. Mean effective gain.

Table 2. Comparison of the proposed antenna with the literature.

Reference	Operating Frequencies (GHz)	Antenna Size (λ)	Isolation (dB)	Gain (dBi)	Fractional Bandwidth (%)	Efficiency (%)	Feed Type
[33]	4.9	2.28 × 0.73 × 0.13	25	6.2	5	–	Microstrip
[34]	3.22–3.97, 4.95–5.51	0.86 × 0.86 × 0.12	18, 20	5.2, 5.5	4.9, 2.3	94	Microstrip
[35]	5.71–8.2, 7.57–7.95	1.53 × 1.53 × 0.12	20, 15	−1.9, 3.8	34.8, 4.5	–	Microstrip
[36]	3.40–4.13	1.13 × 1.13 × 0.19	14	8.1	19.4	91	Probe
[37]	3.50–5.10	1.43 × 1.43 × 0.36	–	8.5	46	89	Probe
[38]	4.56–9.96	1.5 × 1.5 × 0.39	23	–	73.9	–	Trapezoidal Patch
[39]	4.33–7.02	2.6 × 2.6 × 0.36	–	–	48	90	Probe
Proposed Antenna	2.86–3.48 3.67–4.25	0.66 × 0.66 × 0.12	20, 26	5.8, 6.2	18.7, 14.6	88.6, 90.2	Aperture Couple

6. Conclusions

A dielectric resonator-based four-port MIMO antenna is presented. Optimized electromagnetic coupling of fields emanating from patch resonators and fed to the dielectric material induces the dual-mode operation of the antenna. The $TE_{01\delta}$ and $TE_{10\delta}$ modes were excited to achieve dual resonances for target applications. The antenna provides a high gain of 5.8 and 6.2 dBi at 3.3 GHz and 3.9 GHz, respectively. The presented diversity performance of the antenna meets the desired communication applications.

Author Contributions: Conceptualization, T.U. and U.P.; resources, U.P.; writing—review and editing, T.U. All authors have read and agreed to the published version of the manuscript.

Funding: This research received no external funding.

Data Availability Statement: Not applicable.

Conflicts of Interest: The authors declare no conflict of interest.

References

1. Sharawi, M.S. Printed multi-band MIMO antenna systems and their performance metrics [wireless corner]. *IEEE Antennas Propag. Mag.* **2013**, *55*, 218–232. [CrossRef]
2. Karaboikis, M.P.; Papamichael, V.C.; Tsachtsiris, G.F.; Soras, C.F.; Makios, V.T. Integrating compact printed antennas onto small diversity/MIMO terminals. *IEEE Trans. Antennas Propag.* **2008**, *56*, 2067–2078. [CrossRef]
3. Huang, H.; Li, X.; Liu, Y. A low-profile, dual-polarized patch antenna for 5G MIMO application. *IEEE Trans. Antennas Propag.* **2018**, *67*, 1275–1279. [CrossRef]
4. Li, Y.; Zhao, Z.; Tang, Z.; Yin, Y. Differentially fed, dual-band dual-polarized filtering antenna with high selectivity for 5G sub-6 GHz base station applications. *IEEE Trans. Antennas Propag.* **2019**, *68*, 3231–3236. [CrossRef]
5. Le Thi, C.H.; Ta, S.X.; Nguyen, X.Q.; Nguyen, K.K.; Dao-Ngoc, C. Design of compact broadband dual-polarized antenna for 5G applications. *Int. J. RF Microw. Comput. Aided Eng.* **2021**, *31*, e22615. [CrossRef]
6. Chouhan, S.; Panda, D.K.; Kushwah, V.S.; Mishra, P.K. Octagonal-shaped wideband MIMO antenna for human interface device and S-band application. *Int. J. Microw. Wirel. Technol.* **2019**, *11*, 287–296. [CrossRef]
7. Yadav, S.K.; Kaur, A.; Khanna, R. Compact Rack Shaped MIMO Dielectric Resonator Antenna with Improved Axial Ratio for UWB Applications. *Wirel. Pers. Commun.* **2021**, *117*, 591–606. [CrossRef]
8. Girjashankar, P.R.; Upadhyaya, T. Substrate integrated waveguide fed dual band quad-elements rectangular dielectric resonator MIMO antenna for millimeter wave 5G wireless communication systems. *AEU-Int. J. Electron. Commun.* **2021**, *137*, 153821. [CrossRef]
9. Anuar, S.U.; Jamaluddin, M.H.; Din, J.; Kamardin, K.; Dahri, M.H.; Idris, I.H. Triple band MIMO dielectric resonator antenna for LTE applications. *AEU-Int. J. Electron. Commun.* **2020**, *118*, 153172. [CrossRef]
10. Dwivedi, A.K.; Sharma, A.; Singh, A.K.; Singh, V. Circularly polarized quad-port MIMO dielectric resonator antenna with beam tilting feature for vehicular communication. *IETE Tech. Rev.* **2020**, *39*, 389–401. [CrossRef]
11. Alibakhshikenari, M.; Babaeian, F.; Virdee, B.S.; Aïssa, S.; Azpilicueta, L.; See, C.H.; Althuwayb, A.A.; Huynen, I.; Abd-Alhameed, R.A.; Falcone, F.; et al. A comprehensive survey on "Various decoupling mechanisms with focus on metamaterial and metasurface principles applicable to SAR and MIMO antenna systems". *IEEE Access* **2020**, *8*, 192965–193004. [CrossRef]
12. Wang, Z.; Li, C.; Wu, Q.; Yin, Y. A metasurface-based low-profile array decoupling technology to enhance isolation in MIMO antenna systems. *IEEE Access* **2020**, *8*, 125565–125575. [CrossRef]
13. Bhattacharjee, A.; Karmakar, A.; Saha, A.; Bhattacharya, D. Design of a compact UWB MIMO-diversity antenna incorporating fractal inspired isolation structure with band notch characteristics. *Microw. Opt. Technol. Lett.* **2021**, *63*, 2597–2605. [CrossRef]
14. Yang, Z.; Xiao, J.; Ye, Q. Enhancing MIMO antenna isolation characteristic by manipulating the propagation of surface wave. *IEEE Access* **2020**, *8*, 115572–115581. [CrossRef]
15. Tang, J.; Faraz, F.; Chen, X.; Zhang, Q.; Li, Q.; Li, Y.; Zhang, S. A metasurface superstrate for mutual coupling reduction of large antenna arrays. *IEEE Access* **2020**, *8*, 126859–126867. [CrossRef]
16. Deng, J.Y.; Li, J.Y.; Guo, L.X. Decoupling of a three-port MIMO antenna with different impedances using reactively loaded dummy elements. *IEEE Antennas Wirel. Propag. Lett.* **2018**, *17*, 430–433. [CrossRef]
17. Roy, S.; Chakraborty, U. Mutual coupling reduction in a multi-band MIMO antenna using meta-inspired decoupling network. *Wirel. Pers. Commun.* **2020**, *114*, 3231–3246. [CrossRef]
18. Moussa, K.H.; Amar, A.S.; Mabrouk, M.; Mohamed, H.G. Slotted E-Shaped Meta-Material Decoupling Slab for Densely Packed MIMO Antenna Arrays. *Micromachines* **2021**, *12*, 873. [CrossRef]
19. Li, M.; Cheung, S. A novel calculation-based parasitic decoupling technique for increasing isolation in multiple-element MIMO antenna arrays. *IEEE Trans. Veh. Technol.* **2020**, *70*, 446–458. [CrossRef]
20. Ding, C.F.; Zhang, X.Y.; Xue, C.D. Novel pattern-diversity-based decoupling method and its application to multielement MIMO antenna. *IEEE Trans. Antennas Propag.* **2018**, *66*, 4976–4985. [CrossRef]

21. Kumar, S.; Nandan, D.; Srivastava, K.; Kumar, S.; Singh, H.; Marey, M.; Mostafa, H.; Kanaujia, B.K. Wideband circularly polarized textile MIMO antenna for wearable applications. *IEEE Access* **2021**, *9*, 108601–108613. [CrossRef]
22. Huang, J.; Dong, G.; Cai, J.; Li, H.; Liu, G. A quad-port dual-band MIMO antenna array for 5G smartphone applications. *Electronics* **2021**, *10*, 542. [CrossRef]
23. Pant, A.; Singh, M.; Parihar, M.S. A frequency reconfigurable/switchable MIMO antenna for LTE and early 5G applications. *AEU-Int. J. Electron. Commun.* **2021**, *131*, 153638. [CrossRef]
24. El Hadri, D.; Zakriti, A.; Zugari, A.; El Ouahabi, M.; El Aoufi, J. High isolation and ideal correlation using spatial diversity in a compact MIMO antenna for fifth-generation applications. *Int. J. Antennas Propag.* **2020**, *2020*, 2740920. [CrossRef]
25. Jin, X.; Qiu, Y.; Wu, D.; Yu, G.; Guo, R.; Wu, G.; Zhu, M.; Zhou, H.M. A Low-Profile Dual-Polarized MIMO Antenna with an AMC Surface for WLAN Applications. *Int. J. Antennas Propag.* **2021**, *2021*, 9218255. [CrossRef]
26. Laxman, P.; Jain, A. Circularly Polarized Wideband Fabric Stealth Multiple-Input Multiple-Output Antenna for Ultrawideband Applications Useful for Wireless Systems Wearable on Garments. *Int. J. Antennas Propag.* **2021**, *2021*, 1426680. [CrossRef]
27. Daghari, M.; Essid, C.; Sakli, H. Muli-UWB Antenna System Design for 5G Wireless Applications with Diversity. *Wirel. Commun. Mob. Comput.* **2021**, *2021*, 9966581. [CrossRef]
28. Upadhyaya, T.; Park, I.; Pandey, R.; Patel, U.; Pandya, K.; Desai, A.; Pabari, J.; Byun, G.; Kosta, Y. Aperture-Fed Quad-Port Dual-Band Dielectric Resonator-MIMO Antenna for Sub-6 GHz 5G and WLAN Application. *Int. J. Antennas Propag.* **2022**, *2022*, 4136347. [CrossRef]
29. Ali, A.; Tong, J.; Iqbal, J.; Illahi, U.; Rauf, A.; Rehman, S.U.; Ali, H.; Qadir, M.M.; Khan, M.A.; Ghoniem, R.M. Mutual Coupling Reduction through Defected Ground Structure in Circularly Polarized, Dielectric Resonator-Based MIMO Antennas for Sub-6 GHz 5G Applications. *Micromachines* **2022**, *13*, 1082. [CrossRef]
30. Singhwal, S.S.; Kanaujia, B.K.; Singh, A.; Kishor, J.; Matekovits, L. Multiple input multiple output dielectric resonator antenna with circular polarized adaptability for 5G applications. *J. Electromagn. Waves Appl.* **2020**, *34*, 1180–1194. [CrossRef]
31. Iqbal, A.; Nasir, J.; Qureshi, M.B.; Khan, A.A.; Rehman, J.U.; Rahman, H.U.; Fayyaz, M.A.; Nawaz, R. A CPW fed quad-port MIMO DRA for sub-6 GHz 5G applications. *PLoS ONE* **2022**, *17*, e0268867. [CrossRef] [PubMed]
32. Roshani, S.; Yahya, S.I.; Alameri, B.M.; Mezaal, Y.S.; Liu, L.W.; Roshani, S. Filtering Power Divider Design Using Resonant LC Branches for 5G Low-Band Applications. *Sustainability* **2022**, *14*, 12291. [CrossRef]
33. Sarkar, G.A.; Ballav, S.; Chatterjee, A.; Ranjit, S.; Parui, S.K. Four element MIMO DRA with high isolation for WLAN applications. *Prog. Electromagn. Res. Lett.* **2019**, *84*, 99–106. [CrossRef]
34. Dwivedi, A.K.; Sharma, A.; Singh, A.K.; Singh, V. Design of dual band four port circularly polarized MIMO DRA for WLAN/WiMAX applications. *J. Electromagn. Waves Appl.* **2020**, *34*, 1990–2009. [CrossRef]
35. Varshney, G.; Singh, R.; Pandey, V.S.; Yaduvanshi, R.S. Circularly polarized two-port MIMO dielectric resonator antenna. *Prog. Electromagn. Res. M* **2020**, *91*, 19–28. [CrossRef]
36. Fakhte, S.; Oraizi, H. Compact uniaxial anisotropic dielectric resonator antenna operating at higher order radiating mode. *Electron. Lett.* **2016**, *52*, 1579–1580. [CrossRef]
37. Mukherjee, B.; Patel, P.; Mukherjee, J. Hemispherical dielectric resonator antenna based on apollonian gasket of circles—A fractal approach. *IEEE Trans. Antennas Propag.* **2013**, *62*, 40–47. [CrossRef]
38. Sharma, A.; Biswas, A. Wideband multiple-input–multiple-output dielectric resonator antenna. *IET Microw. Antennas Propag.* **2017**, *11*, 496–502. [CrossRef]
39. Maity, S.; Gupta, B. Experimental investigations on wideband triangular dielectric resonator antenna. *IEEE Trans. Antennas Propag.* **2016**, *64*, 5483–5486. [CrossRef]

MDPI
St. Alban-Anlage 66
4052 Basel
Switzerland
www.mdpi.com

Micromachines Editorial Office
E-mail: micromachines@mdpi.com
www.mdpi.com/journal/micromachines

Disclaimer/Publisher's Note: The statements, opinions and data contained in all publications are solely those of the individual author(s) and contributor(s) and not of MDPI and/or the editor(s). MDPI and/or the editor(s) disclaim responsibility for any injury to people or property resulting from any ideas, methods, instructions or products referred to in the content.

www.ingramcontent.com/pod-product-compliance
Lightning Source LLC
Chambersburg PA
CBHW061617210926
51850CB00009B/2352
* 9 7 8 3 7 2 5 8 0 4 5 0 4 *